Eco-Architecture IX

WIT*PRESS*

WIT Press publishes leading books in Science and Technology.
Visit our website for the current list of titles.
www.witpress.com

WIT*eLibrary*

Home of the Transactions of the Wessex Institute.
Papers contained in this volume are archived in the WIT eLibrary in volume 210 of WIT Transactions on the Built Environment (ISSN 1743-3509).
The WIT eLibrary provides the international scientific community with immediate and permanent access to individual papers presented at WIT conferences.
Visit the WIT eLibrary at www.witpress.com.

NINTH INTERNATIONAL CONFERENCE ON
HARMONISATION BETWEEN
ARCHITECTURE AND NATURE

Eco-Architecture 2022

CONFERENCE CHAIRMEN

Pilar Chias
University of Alcala de Henares, Spain

Santiago Hernández
University of A Coruña, Spain
Member of WIT Board of Directors

INTERNATIONAL SCIENTIFIC ADVISORY COMMITTEE

ORGANISED BY

Wessex Institute, UK
University of Alcala de Henares, Spain

SPONSORED BY

WIT Transactions on the Built Environment
International Journal of Environmental Impacts
International Journal of Energy Production and Management

WIT Transactions

Wessex Institute
Ashurst Lodge, Ashurst
Southampton SO40 7AA, UK

We would like to express thanks to all the conference Chairs and members of the International Scientific Advisory Committees for their efforts during the 2022 conference season.

Conference Chairs

International Scientific Advisory Committee Members 2022

Eco-Architecture IX

Editors

Pilar Chias
University of Alcala de Henares, Spain

Santiago Hernández
University of A Coruña, Spain
Member of WIT Board of Directors

WIT*PRESS* Southampton, Boston

Editors:

Pilar Chias
University of Alcala de Henares, Spain

Santiago Hernández
University of A Coruña, Spain
Member of WIT Board of Directors

Published by

WIT Press
Ashurst Lodge, Ashurst, Southampton, SO40 7AA, UK
Tel: 44 (0) 238 029 3223; Fax: 44 (0) 238 029 2853
E-Mail: witpress@witpress.com
http://www.witpress.com

For USA, Canada and Mexico

Computational Mechanics International Inc
25 Bridge Street, Billerica, MA 01821, USA
Tel: 978 667 5841; Fax: 978 667 7582
E-Mail: infousa@witpress.com
http://www.witpress.com

British Library Cataloguing-in-Publication Data

A Catalogue record for this book is available
from the British Library

ISBN: 978-1-78466-473-2
eISBN: 978-1-78466-474-9
ISSN: 1746-4498 (print)
ISSN: 1743-3509 (on-line)

The texts of the papers in this volume were set individually by the authors or under their supervision. Only minor corrections to the text may have been carried out by the publisher.

Preface

This volume contains a selection of papers among those presented at the ninth International Conference on Harmonisation between Architecture and Nature, organised by the Wessex Institute. This conference series started in the New Forest, home of the Wessex Institute in 2006, and then in the Algarve (Portugal) in 2008, A Coruña (Spain) in 2010, the island of Kos (Greece) in 2012, Siena (Italy) in 2014, Alicante (Spain) in 2016, again in the New Forest in 2018 and the city of A Coruña was also intended for the edition in 2020 but the event was finally held online due to the COVID-19 pandemic.

In 2022 the Wessex Institute and the University of Alcalá de Henares (Spain) organized the conference, and the venue selected was Lisbon, the beautiful capital city of Portugal. However, the health crisis created by COVID-19 continuing to reduce people's mobility and the difficult political and military situation in Eastern Europe forced to change again the in person conference into an online forum.

The objective of this Conference is to review the challenges and new opportunities for contemporary architecture. These are a result of advances in design and new building technologies, as well as the development of new materials. Contemporary architecture is at the threshold of a new stage of evolution, deeply influenced by the advances in information and computer systems and the development of new materials and products, as well as construction processes.

Emphasis is placed on the minimum use of energy at each stage of the building process, the design by passive systems, life cycle assessment, resources optimisation and rehabilitation.

Never before in history have architects and engineers had such a range of new processes and products open to them. In spite of that, the construction industry lags behind all others in taking advantage of a wide variety of new technologies. This is understandable, due to the inherent complexity and uniqueness of each architectural project. Advances in computer and information systems, offers the possibility of developing new architectural forms, construction products and building technologies.

Eco-Architecture always aims at designs that adapt to the environment, learning from nature and it is linked also to sustainable or "green" design. Adopting traditional architectural concepts can help to achieve this, such as, for instance, a philosophy promoting harmonious equilibrium among nature, buildings and people or courtyards designed to perform as passive energy systems apart from serving the community's cultural needs. Eco-Architecture also promotes the idea of using the minimum amount of energy at each stage of the building process, including the amounts required during the extraction and transportation of materials, their fabrication, assembly, building erection, maintenance and eventual future recycling.

Contributions to Eco-Architecture 2022 included in this book correspond to many and varied topics. Some deal with the application of circular economy approach to eco-architecture, others are related to energy performance of buildings and describe the influence of vacuum insulating systems, the advantages of mimicking nature to diminish energy consumption, the characterization of the thermal mass in buildings or the inclusion of nano materials in retroffiting techniques. Some papers describe research on social factors as occupant's behaviour, the influence of design in restaurants service, the implication of city residents in urban gardens or the need to combine nature and technology to secure healthy buildings.

Sustainability of cities is one of main topics of the conference, and presentations described the use of biophilic buildings design, the evaluation of the climate change in residential neighbourhoods. The materials used in architecture is the subject of some papers included in this book and their approaches go from the renewed use of compressed earth blocks to the use of organic materials, the use of engineering wood and the capabilities of graphene in architecture.

The situation of current architectural design and the perspectives for the future was discussed by some authors in papers about the common absence of details in modern architecture, the need for user-centred design, the changes in topography produced by buildings, the chances for future floating cities

The application examples or the papers corresponded to very different locations and design constraints, ranging from arid zones to tropical environments and from new urban developments to adaptive reuse of historical sites.

The conference is deliberately international in scope and seeks to provide an opportunity for sharing experience and potential solutions to common problems. In that regards, the participants, belonged to several countries and continents and the scheme of the online event arranged by the Wessex Institute made it easy for the participants to access to the papers accepted after a rigorous peer review procedure. The Q&A sessions provided a friendly atmosphere and were very useful for interaction between authors and delegates and interchange of new ideas and initiatives.

We consider this volume to be a valuable addition to the existing literature on ecological approaches in architecture and will, therefore, be of interest to any researcher working in this scientific field.

These papers, like others presented at Wessex Institute conferences, are referenced by CrossRef and appear regularly in suitable reviews, publications and databases, including referencing and abstracting services. They are also archived online in the WIT eLibrary (http://www.witpress.com/elibrary) where they are permanently available in Open Access format to the international scientific community.

The Editors would like to thank the authors for their contributions, as well as the members of the International Scientific Advisory Community of the Conference and the reviewers of the papers for their invaluable help in ensuring the high scientific level of the event.

The Editors, 2022

Contents

SECTION 1
CIRCULAR ECONOMY

SUSTAINABLE MODEL FOR ARCHITECTURE AND RETAILING ENVIRONMENTS BASED ON THE CIRCULAR ECONOMY CONCEPT

AYAT SLEIMAN
Lawrence Technological University, USA

ABSTRACT

A recent sustainability-based economic model, the Circular Economy, is examined as a catalyst for a potential new retail model focused on an integrated sustainable system providing residential building and renovation products sales. The premise of the Circular Economy is the argument of utilizing resources continuously by extracting the maximum value from them, then recovering and regenerating products and materials at the end of each service life. The research presents a qualitative study of the Circular Economy concept as an innovation outgrowth for the built environment and retailing facilities. The work proves the Circular Economy theory's necessity as applied to the retail industry, providing a new programmatic retail-based concept with a practical development as an architectural design demonstration project. This project demonstrates how through new retail programmatic synergies, the Circular Economy would be the basis for a new retail model. The application is shown through an urban redevelopment proposal of a former industrial building in a declining North American city that demonstrates Circular Economy issues in the site selection and building design.
Keywords: waste, circular economy, collect, fix, sell, residential, retail, regulations.

1 INTRODUCTION

By the end of the second decade of the 21st century, sustainability and green building materials have become the leading criteria for the architecture practice in North America. Despite the vast expansion of green building materials in the U.S., their availability to consumers in the residential home improvement and renovation arena is minimal, even though Americans spend more than $400 billion a year on residential renovations and repairs [1].

More than 60% of consumers want more sustainable product choices [2]. However, with limited green building retail material stores, premium prices, and a lack of awareness of the benefits of green products, sustainability is facing a major deficit in the field of residential home improvement. The current home improvement retail models are operating on a linear economy aspect of "take, make, and waste" [3]. The absence of sustainable home improvement stores from the reach of the American society while the linear economy retail models are monopolizing the residential market leads to overusing natural resources and producing more landfill waste.

Waste is one of the most significant challenges to civilization. More than 2.1 billion tons of waste are produced every year, and only 16% of them are being recycled [4] – the rest go to landfills. According to the Verisk Maplecroft report, the U.S. represents 4% of the world's population, but it produces 12% of global waste [5]. Americans are considered the originators of the concept "throwaway society", as described by the sociologist Vance Packard, due to their acquiring habit that is attributed to cultural hegemony of consumption and the unlimited regenerative styles of a variety of commodities that change with fads or fashion. As a result, U.S. consumers replace their older items with newer ones frequently, and they produce unsustainable amounts of waste. Products in the home improvement arena fall within this condition. In fact, construction remodeling materials account for 22% of that waste according to the U.S. EPA [6].

WIT Transactions on The Built Environment, Vol 210, © 2022 WIT Press
www.witpress.com, ISSN 1743-3509 (on-line)
doi:10.2495/ARC220011

The closed-loop regenerative approach of the Circular Economy concept, which centers on "take, make, and reuse" instead of "take, make, and waste", is a promising solution for the issue of waste and the lack of sustainable products. It also limits the extraction of the primary resources while supporting economic growth and generating new green collar jobs through upcycling and recycling [7].

This study examined the application of the Circular Economy theory in an architectural, programmatic design proposal. The research includes a qualitative study with inductive and deductive methodologies examining the Circular Economy innovation outgrowth for the built environment and retailing facilities. The first part of this research explores the problem, the available solutions and the pros and cons of the results and concludes that the Circular Economy concept is the solution to the dilemma facing the current retail model, and that it is necessary for architects to change the design of the current model of retailing facilities. The second part presents a specialized theoretical concept and proposes a demonstration project that uses Circular Economy principles to create a hybrid of current retail models and interactive social spaces in one facility. The new model gives consumers access to sustainable building products in an attractive, safe, and communal environment and includes repurposed and recycled goods.

2 THE IMPACT OF CONSUMERISM ON WASTE

The Industrial Revolution fostered mass consumption and production while extracting natural resources extravagantly [8]. This, combined with marketing strategies, serves to booster the American dream; regardless of their purchasing power, consumers acquire more items than they need. "With less than 5 percent of the world's population, the U.S. uses one-third of the world's paper, a quarter of the world's oil, 23 percent of the coal, 27 percent of the aluminum, and 19 percent of the copper". People in the U.S. waste more than the environment can afford to process [9], and they throw away tons of items that have not lost their full life cycle, items that can be recovered, reused or resold [10].

Consumerism also played an essential role in the architecture of retail facilities in the U.S., affecting the production of buildings' form developing a consumerist type of architecture like shopping malls, department stores, chain food restaurants, and many other. In the last decade, the U.S. has added more than 0.9 billion square feet of vacant retail spaces [11]. Many of those vacant structures are on the demolishing list [12], generating more waste that goes to landfill.

It is very disturbing to see the quantities of valuable products that have been sent to waste without consideration of its residual value. However, that is because there haven't been any serious economic attempts to guide the fate of inadequate items or to urge the economy toward generating new resources out of waste.

3 THE CIRCULAR ECONOMY SOLUTIONS

Experts predict that the Circular Economy will drive the fourth economic revolution [13]. Many countries have already adopted the Circular Economy measures into their economic system. The British Standards Institution (BSI), for example, developed and launched its first Circular Economy standard, and the European Commission approved an action plan to boost Europe's transition to Circular Economy.

The concept of circularity can be applied to many industries. In fact, major worldwide companies have already adopted the Circular Economy concept into their products in industries like high tech, automotive, textile, agriculture, food, furniture, biochemical, architecture and construction, business, materials, development institutions and others [14].

3.1 Circular Economy overview

Circular Economy is a systematic approach to economic development designed to benefit businesses, society, and the environment. A Circular Economy is regenerative by design and aims to decouple growth from the consumption of finite resources gradually [15]. It also utilizes closed-loop systems to divert materials from landfills – recovering or restoring products, components, and materials by reusing, repairing, remanufacturing, or recycling. Closed-loop recycling can increase the lifespan of a product and optimally become infinitely regenerative [16]. The Circular Economy "model of production represents an unprecedented opportunity to reduce costs, embodied carbon, and strain on natural resources and ecosystems" [17].

The Ellen MacArthur Foundation illustrates the Circular Economy concept in a "butterfly" diagram. The right side illustrates the technical cycle and closing the loops of resources by circularity strategies such as reuse, refurbishment, and recycling. The left side of the diagram shows the biological cycle and the loops and cascades assuring the sustainable management of biological resources and creating renewable flows and stocks. The aim of this model is to minimize the extraction of raw materials and waste generation [18] (Fig. 1).

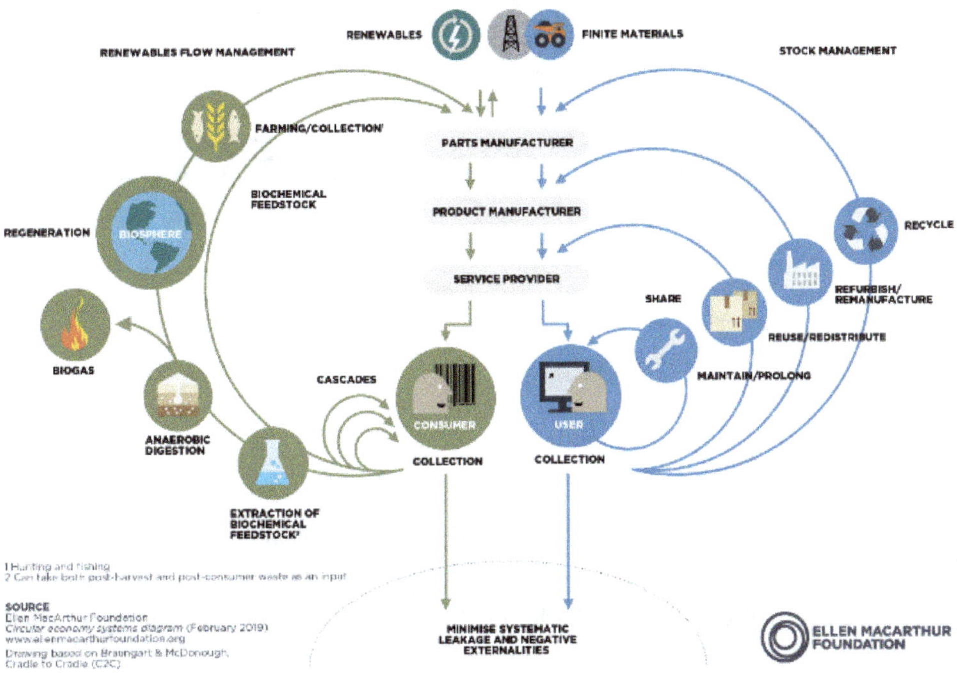

Figure 1: The butterfly diagram – Visualizing the Circular Economy.

3.2 Circular Economy in the built environment

The largest part of our daily experiences is integrated with the built environment. The quality of the building that surrounds us influences our emotions and actions. Beyond the force of emotions that the built environment provides, there are technical aspects, such as deconstruction, which are becoming a deep interest for the architecture community [19].

The Ellen MacArthur Foundation diagram is highly relevant to the built environment. The right side guides the construction process and enforces the disassemble design principles, and the left side handles the after-use materials and factors them into the repurposing and recycling processes that generate new resources out of waste.

Architects and designers can help reduce waste by applying Circular Economy characteristics to buildings and design buildings to be more adjustable, more extensive, and service multiple users, reducing the demand for new buildings. In addition, architects can close the loop for a building's construction and performance by adopting refurbished materials into designs, utilizing restored and recycled goods from the previous program, and designing new products with the intent of prolonged uses [20] (Fig. 2).

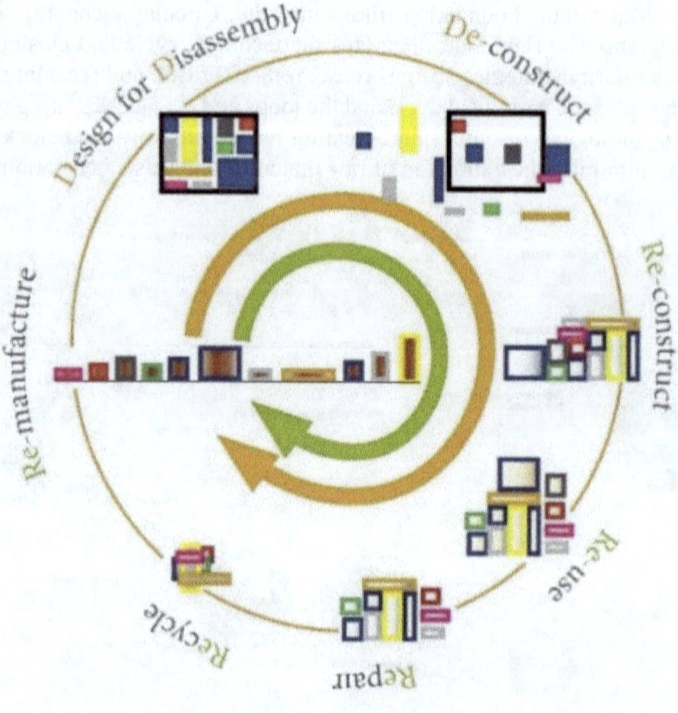

Figure 2: Circular Economy building principles

There are two different approaches for the Circular Economy in the built environment; one involves an existing building, and the second is for new buildings. Existing building principles emphasize extending the life cycle of the building; minimizing resource input, waste, emissions, water, and energy leakage while slowing, closing, and narrowing energy and material loops; and avoiding disposal and loss of economic and ecological value. New building principles emphasize robust and flexible design, as well as guidelines for disassembly, reuse possibilities, repair diagrams, and setups to remanufacture materials or recycle them [21].

The Circular Economy building principles direct architects and designers to approach the built environment on a case-by-case basis to minimize consumption and reduce construction waste by continually reusing materials. The built environment practitioners should reconcile this alternative economic system with their designs, processes, and business models [22].

3.3 Circular Economy business model

In the process of applying the Circular Economy to architecture design and construction, it became clear that is tremendously difficult to close the loop on sustainable building – especially for homeowners – because of a lack of access to sustainable products in the retail home improvement arena [23]. Most of the home improvement stores in the U.S. don't widely practice the Circular Economy business model yet; this movement toward circularity is still in its infancy.

The theoretical concept of the Circular Economy vigorously emphasizes the idea of re-use. A business model based on the re-use theory already exists in businesses like thrift stores, second-hand stores, flea markets, antique shops, pawn shops, consignment stores, record stores, etc. Although re-use businesses make up a small percentage of retail businesses, the demand for reused products is growing, and re-use stores exist all over North America. According to IBIS World, the re-commerce industry is growing at a rate of 2.3% [24], but this at this rate the transformation to the Circular Economy is too far on the horizon.

In addition, re-commerce businesses are mostly independent and localized, and their inventories are in bulks and not specialized into standard products. Most importantly, the managing system capitalizes on the Collect/Sell-As-Is business strategy. Some items sold in the re-use shopping arena are not ready for immediate use due to wear and tear or defects. In many cases, neither the re-commerce shopping arenas nor the customers are willing to fix them. There are guidelines that control the collection and distribution of re-used goods or government rules that support environmental concerns and minimize excessive waste.

In contrast, current home improvement retail stores under the Linear Economy business model are well organized, specialized, and accessible to many homeowners. In fact, the 6.3% growth rate in this industry in the U.S. shows that there is a demand for home improvement products [25].

There is no established chain of green building materials stores around the U.S. that is dedicated to homeowners. Therefore, there is a need for a new retail business model in the home improvement industry that is chaperoned by the Circular Economy guidelines.

4 CIRCULAR ECONOMY RETAIL BUSINESS MODEL

The theory of Circular Economy should change the way architects design retailing facilities. Architects and designers need to add the layer of circularity into their plan so that the design acknowledges and allows the operation of the space under the Circular Economy concept. Combining the Circular Economy architecture characteristics with the Circular Economy business characteristics in the residential home improvement arena creates a new approach: the Circular Economy retail business model [26], which utilizes reused and recycled material all in one place while minimizing waste. Unlike the current re-commerce stores that operate in a one-way system of "Collect-Sell", the new model operates on a closed loop strategy of "Collect-Fix-Sell" (Fig. 3).

4.1 New synergies in retailing

The resulting demonstration project operates the Collect-Fix-Sell system in a single facility as an alternative to or hybrid of current retail models [27]. In this innovative design, consumers will have access to multiple services that revolve into one continuous loop of generating new sources, making profits, and sustaining an environmentally friendly attitude. Buyers and sellers will have access to collection storage facilities, workshops, retail sales,

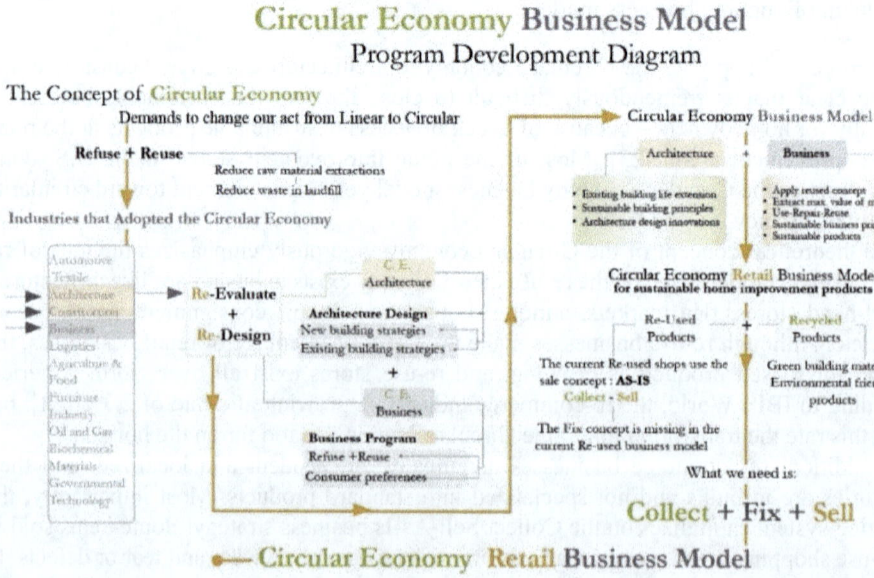

Figure 3: Circular Economy business model program development diagram.

and interactive social spaces. Users will be able to buy, fix and sell sustainable home improvement goods in an attractive, safe, communal, and comfortable environment (Fig. 4).

A functional facility that combines storage, light industrial work and a market will promote the concept of up-cycling and recycling products and raise the discourse of sustainable awareness, solidarity, and responsibility among the public.

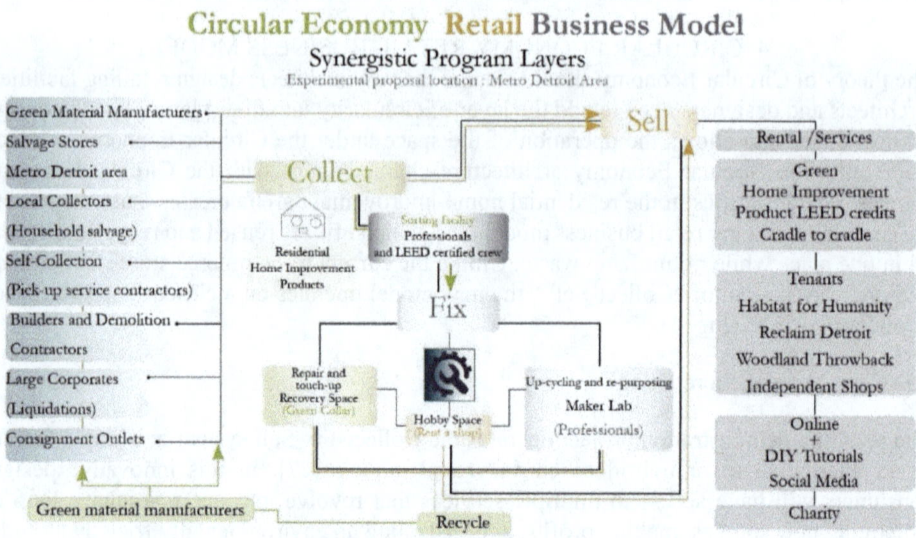

Figure 4: Circular Economy retail business model flowchart diagram.

4.2 The Circular Economy retail business model program layers

The proposed business model distribution plan starts in the collection facility "Collect". This department is responsible for compiling all salvage architecture items and green building materials. The collected items are assessed and sorted by professionals to determine their eligibility for immediate sale or repair before they reach the public. Example of items collected include reclaimed lumber, building materials, and other architecture details. The collection space will require a loading dock, sorting facility and storage.

Items from the "Collect" area will be send to one of the three "Fix" facility departments: Recovery, Maker Lab or Hobby Space. These specialized departments assess the items and extend their life span through repairing, upcycling, or repurposing instead of throwing them away. The "Sell" facility is the retail area, where public can access a wide variety of recycled and upcycled green home improvement items and materials.

4.3 The Circular Economy retail business model application

Demonstrating the Circular Economy theory in a practical retail business model required a suitable building and a site with a supportive urban fabric. The process of qualifying a site for the new retail business model began with an analysis of the causes of the retail apocalypse in North America [28], the changes in consumer spending habits [29], the availability of products that could be reused, and existing abandoned buildings [30] capable of accommodating the programmatic elements for the new retail store. (Fig. 5).

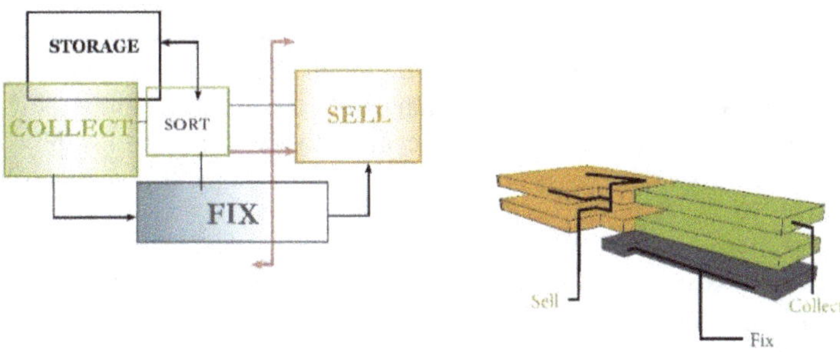

Figure 5: Program major spaces flowchart diagram.

4.3.1 Existing building selection

There are many abandoned and vacant buildings in North America due to the retail apocalypse [30]. This research tested three different building sites with structures that allowed flexibility in their plan for a multi-purpose program and their ability to attract foot traffic, capitalize on interest in re-commerce products, support urban revival, and improve commercial zones.

The best option for the application of Circular Economy principles was the old Apple Thorn Valley Slaughterhouse property located in northeast Detroit, Michigan (Fig. 6). It was built in 1937 near the Eastern Market, which is a vibrant site sponsored by the revitalization plan of the city of Detroit. The building has been vacant for nearly two decades [31]. In addition to the historical value of the building, the structure has had changes to its program and form throughout the years, making its floor plan flexible.

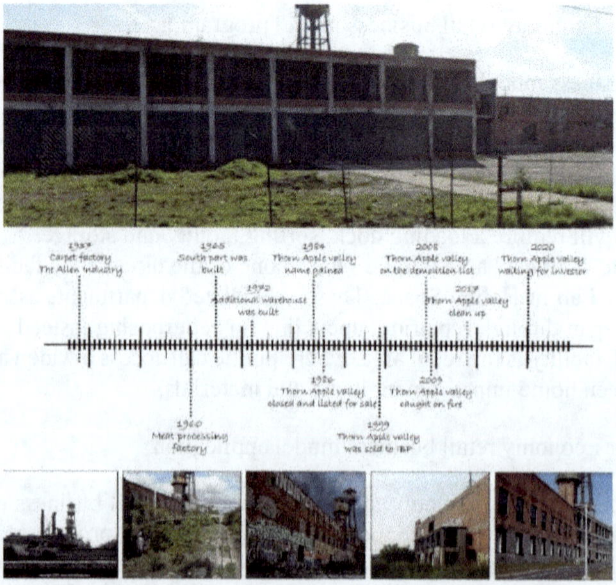

Figure 6: Building site with historical timeline.

4.3.2 Market analysis
This building is in the center of an emerging plan for the City of Detroit, which aims to enhance the business growth in the area and attract high foot traffic to the nearby market and a variety of attractions [32]. The vibrant location capitalizes on the demographic changes and the interest in Gen Z who show a wide embracement to the re-use philosophy, and they are willing to pay more for sustainable products (73%) compared to Millennials (68%) [33].

4.3.3 Building analysis
The building is a two-story structure with an additional walkout lower level that is located in an industrial zone and features a 2.13-acre lot with 1100 parking spots on site and near the building. The top two floors provide an approximate 115,000 square feet; the lower level provides approximately 28,000 square feet that features a 471 linear feet access to Dequindre Cut, a public green walkway with variety of attractions [34]. A potential opening on the southwest wall could be utilized as a corridor connection to drive the foot traffic through the building from the market on the southwest corner to Dequindre Cut in the northeast. In addition, the building across from the subject property is also vacant and could potentially serve the expansion program in the future [35].

Preliminary bioclimatic studies suggest minor changes to the exterior shell to preserve more energy and enhance the natural ventilation system and day lighting performance [36]. Based on the building analysis, the property is highly suitable for the proposed model (Fig. 7).

A closer look at the interior architecture and the program function of the building found an incredible opportunity to enhance the circulation and the flow for the proposed business plan [37]. The plan enhances the form and the function of the exterior facades and the interior flow of the building while maximizing the advantages of the building's orientation, location, circulation, and bioclimatic design strategies (Fig. 8).

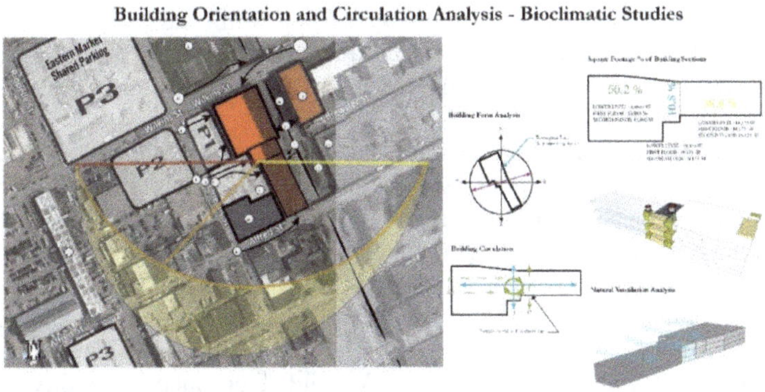

Figure 7: Building site circulation and bioclimatic analysis.

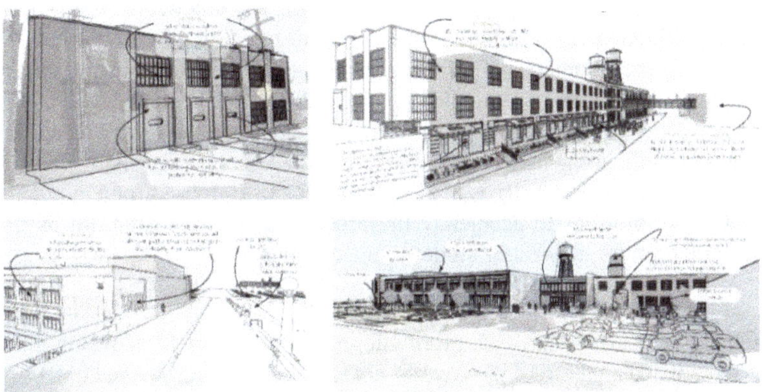

Figure 8: Building design analysis proposal.

Figure 9: Circular Economy retail business rendering.

The former Thorn Apple Valley Slaughterhouse building will be transformed from an abandoned site into a vibrant destination with retail shops, food venues, exhibitions, storage, workshop spaces, and a market for sustainable home improvement building products. The project preserves the energy of the original building while adapting it into an energy-conserving building and revitalizing a Detroit landmark. The retail business capitalizes on the existing high foot traffic levels and consumers that come to the Eastern Market for locally made products (Fig. 9).

In conclusion, this research addressed the issue of excessive waste in the U.S. and the lack of sustainable product choices in the residential home improvement arena despite the rise of sustainability movement and consumer preferences for reused products. This paper described the effects of mass production, consumerism, consumer behavior and retail closures that led to negative impacts on the environment due to massive extraction of natural resources and excessive waste generation from throw-away products and construction residue. This study also examined the concept of the Circular Economy as a promising solution to the issue of waste and sustainable products in the retail market and determined its validity.

Based on this conclusion, this work proposed a theoretical, programmatic concept for a retail business model based on the Circular Economy precepts and presented an architectural design demonstration project; both the architecture and the business framing systems combined were able to prove and present a new guideline for a retail business model for home improvement products that promote recycling and upcycling and presents innovative synergistic shopping relationships. The proposed design and site demonstrate the sustainable practices espoused to customers with storage, workspace, and market areas that emphasize the Circular Economy concept. This example is a guide for developers and designers to invest in and create profitable home improvement retail chain stores.

The U.S. transition to circularity is in its adolescence [38]. It is a problem but also a major opportunity to test solutions to accelerate the process, as this research did by proposing a guideline to transform a linear retail model into a circular one.

REFERENCES

[1] Wedeen, S. & Carlos Martín, C., Remodeling. Joint Center for Housing Studies, 2021. https://www.jchs.harvard.edu/research-areas/remodeling. Accessed on: 17 Mar. 2022.

[2] NielsenIQ, Was 2018 the year of the influential sustainable consumer? 2021. https://nielseniq.com/global/en/insights/analysis/2018/was-2018-the-year-of-the-influential-sustainable-consumer/. Accessed on: 17 Mar. 2022.

[3] Vincent, N., Yoharshvardhan. Contextualizing the circular economy in building design. *Bollywood News*, 2022. https://bankofbollywood.com/?p=679. Accessed on: 17 Mar. 2022.

[4] *The Guardian*, US produces far more waste and recycles far less of it than other developed countries. 2019. https://www.theguardian.com/us-news/2019/jul/02/us-plastic-waste-recycling. Accessed on: 17 Mar. 2022.

[5] Turrentine, J., The United States is the most wasteful country in the world. NRDC, 2022. https://www.nrdc.org/onearth/united-states-most-wasteful-country-world. Accessed on: 17 Mar. 2022.

[6] Reducing waste during home remodeling. Choose to Reuse. https://www.hennepin.us/choose-to-reuse/tips/home-remodeling-reduce-waste. Accessed on: 17 Mar. 2022.

[7] steelconstruction.info, Steel and the circular economy. https://www.steelconstruction.info/Steel_and_the_circular_economy. Accessed on: 17 Mar. 2022.

[8] Khan Academy, The industrial revolution. https://www.khanacademy.org/humanities/big-history-project/acceleration/bhp-acceleration/a/the-industrial-revolution. Accessed on: 26 Apr. 2022.

[9] Schor, J. & Holt, D.B., *The Consumer Society Reader*, eds J. Schor, D.B. Holt & H. Holt, New Press, 2014. https://books.google.at/books?id=IXRzks8UEv0C.

[10] Belle, R., What actually happens to the stuff we donate to thrift stores? MyNorthwest.com, 2019. https://mynorthwest.com/1655464/secondhand-thrift-store-donations-adam-minter/. Accessed on: 26 Apr. 2022.

[11] Leasetome.com, US retail vacancy may increase by a billion sq. ft. in 2020 (or not). https://leasetome.com/blog/us-retail-vacancy-may-increase-by-a-billion-sq-ft-in-2020-or-not/. Accessed on: 26 Apr. 2022.

[12] IBISWorld, Industry market research, reports, and statistics. https://www.ibisworld.com/united-states/market-research-reports/demolition-wrecking-industry/. Accessed on: 26 Apr. 2022.

[13] steelconstruction.info, Steel and the circular economy. https://www.steelconstruction.info/Steel_and_the_circular_economy. Accessed on: 17 Mar. 2022.

[14] HPEFS, IT Asset Management and Lifecyle Solutions and Services, IT asset management and lifecyle solutions and services. https://www.hpe.com/uk/en/services/asset-lifecycle.html. Accessed on: 17 Mar. 2022.

[15] Ellen MacArthur Foundation, The circular economy in detail. https://archive.ellenmacarthurfoundation.org/explore/the-circular-economy-in-detail. Accessed on: 17 Mar. 2022.

[16] Ellen MacArthur Foundation, Circular economy introduction. https://ellenmacarthurfoundation.org/topics/circular-economy-introduction/overview. Accessed on: 17 Mar. 2022.

[17] U.S. Green Building Council, LEED for cities and communities. https://www.usgbc.org/leed/rating-systems/leed-for-cities. Accessed on: 17 Mar. 2022.

[18] Ellen MacArthur Foundation, The butterfly diagram: Visualising the circular economy. How to build a circular economy. https://ellenmacarthurfoundation.org/circular-economy-diagram. Accessed on: 17 Mar. 2022.

[19] TU Delft, Circular economy. https://www.tudelft.nl/en/ceg/about-faculty/departments/engineering-structures/sections-labs/resources-recycling/research-innovation/circular-economy. Accessed on: 26 Apr. 2022.

[20] katch-e, Contextualizing the circular economy in building design. 2017. https://www.katche.eu/contextualizing-circular-economy-building-design/. Accessed on: 17 Mar. 2022.

[21] TU-Delf University, Circular Economy in the Built Environment Course. 2020. https://www.tudelft.nl/en/ceg/about-faculty/departments/engineering-structures/sections-labs/resources-recycling/research-innovation/circular-economy/circularity-in-the-built-environment.

[22] Adey, S., Siufan Adey, Dezeen. https://www.dezeen.com/author/siufan-adey/. Accessed on: 17 Mar. 2022.

[23] IBISWorld, Industry market research, reports, and statistics. http://www.ibisworld.com/united-states/market-research-reports/thrift-stores-industry/. Accessed on: 17 Mar. 2022.

[24] Businesswire.com, GreenPrint survey finds consumers want to buy eco-friendly products, but don't know how to identify them. 2021 https://www.businesswire.com/news/home/20210322005061/en/GreenPrint-Survey-Finds-Consumers-Want-to-Buy-Eco-Friendly-Products-but-Don%E2%80%99t-Know-How-to-Identify-Them. Accessed on: 26 Apr. 2022.

[25] IBISWorld. Industry market research, reports, and statistics. https://www.ibisworld.com/united-states/market-research-reports/home-improvement-stores-industry/. Accessed on: 17 Mar. 2022.
[26] Sleiman, A., A Sustainable model for architecture and retailing environments based on the Circular Economy concept. Thesis, Southfield, Lawrence Technological University, 2020.
[27] Sleiman, A., A sustainable model for architecture and retailing environments based on the Circular Economy concept. Thesis, Southfield, Lawrence Technological University, 2020.
[28] Cnn.com, https://www.cnn.com/2020/01/08/business/macys-store-closures/index.html. Accessed on: 26 Apr. 2022.
[29] Thompson, D., What in the world is causing the retail meltdown of 2017? *The Atlantic*. 2017. https://www.theatlantic.com/business/archive/2017/04/retail-meltdown-of-2017/522384/. Accessed on: 26 Apr. 2022.
[30] Thompson, D., What in the world is causing the retail meltdown of 2017? Futurist.gr., 2017 https://futurist.gr/what-in-the-world-is-causing-the-retail-meltdown-of-2017-by-derek-thompson/. Accessed on: 26 Apr. 2022.
[31] Detroiturbex.Com, Thorn Apple Valley Slaughterhouse. http://www.detroiturbex.com/content/industry/thorn/index.html. Accessed on: 26 Apr. 2022.
[32] City of Detroit, Eastern market. https://detroitmi.gov/departments/planning-and-development-department/neighborhood-plans/central-design-region/eastern-market. Accessed on: 26 Apr. 2022.
[33] Curtin, M., 73 percent of millennials are willing to spend more money on this 1 type of product. Inc., 2018. https://www.inc.com/melanie-curtin/73-percent-of-millennials-are-willing-to-spend-more-money-on-this-1-type-of-product.html. Accessed on: 26 Apr. 2022.
[34] City of Detroit, City of Detroit parcel viewer. https://detroitmi.gov/webapp/city-detroit-parcel-viewer. Accessed on: 27 Apr. 2022.
[35] Sleiman, A., A sustainable model for architecture and retailing environments based on the Circular Economy concept. Thesis, Southfield, Lawrence Technological University, 2020.
[36] Vassigh, S., Özer, E. & Spiegelhalter, T., *Best Practices in Sustainable Building Design*, J. Ross Publishing: Boca Raton, FL, 2013.
[37] City of Detroit, City of Detroit parcel viewer. https://detroitmi.gov/webapp/city-detroit-parcel-viewer. Accessed on: 27 Apr. 2022.
[38] UNEP, North America and the circularity transition. 2020. https://www.unep.org/news-and-stories/speech/north-america-and-circularity-transition. Accessed on: 27 Apr. 2022.

DESIGNING AN URBAN FOOD SYSTEM FOR ACHIEVING CIRCULAR ECONOMY TARGETS: A CONCEPTUAL MODEL

TONG ZOU[1], ALI CHESHMEHZANGI[1,2], AYOTUNDE DAWODU[1] & EUGENIO MANGI[1]
[1]Department of Architecture and Built Environment, Faculty of Science and Engineering, University of Nottingham Ningbo China, China
[2]Network for Education and Research on Peace and Sustainability (NERPS), Hiroshima University, Japan

ABSTRACT

In exploring the strategies for reaching sustainable development goals (SDGs), the importance of circular development has been greatly recognized in sustainable studies and practices. Regarding applying the circularity concept at city scales, the United Nations Economic Commission for Europe emphasizes that transitioning to a circular economy will help city leaders achieve SDGs and other global objectives of climate actions. The food sector is a significant contributor to climate change, biodiversity loss, and land degradation, which can all be resolved via circular economy solutions. There is an urgent need for our current dominant global food systems (FS) to change due to the multiple environmental influences across the whole food value chain (FVC). This calls for the implementation of circular economy principles in FS, especially for urban systems, which account for most of the world's food consumption, loss, and waste; and for most of the world's greenhouse gas emissions. Applying circular economy principles to each stage of FVC offers various opportunities to address FS issues as they enable each actor to identify the parts they can participate in FS transition. Secondly, circular economy strategies such as localizing food chains, food wastes reduction, and reusing raw materials offer solutions to FS sustainability, connecting FS sustainability transition with the transformation of industrial food metabolism from linear to circular. Since cities play a critical role in sustainable transitions from linear to circular, pursuing a circular urban food system (UFS) benefits circular urban metabolism too. To aid the implementation and promotion of circular economy in FS and cities, this study proposes a conceptual model of designing UFS for achieving circular economy targets.

Keywords: circular economy, food system, sustainability, conceptual model, food value chain.

1 INTRODUCTION

With the growing impacts of climate change and other global environmental challenges, the concept of circular economy (CE) as one of the practical strategies toward sustainable development has attracted considerable attention in recent years. It is believed that a circular economy can provide the solutions to current global environmental challenges through a systematic approach by the following three principles [1]:

1. Eliminate and/or design out waste and pollution;
2. Circulate products and materials/keep them in use at their highest value;
3. Regenerate natural system.

At the urban level, a circular city is seen as a promising way to transit from linear to circular urban metabolism. Promoting circular cities' implementation offers pathways to close material loops, reduce and recover material and resources, reduce greenhouse gas emissions and environmental impacts, protect, and improve biodiversity, and reduce social inequities in line with the SDGs [2].

Meanwhile, the food system (FS) discourse also received growing attention in the past ten years as food and agriculture are regarded as the cross-cutting elements for the United

WIT Transactions on The Built Environment, Vol 210, © 2022 WIT Press
www.witpress.com, ISSN 1743-3509 (on-line)
doi:10.2495/ARC220021

Nations 2030 Sustainable Development Goals (SDGs) [3]. FS plays a critical role in realizing multiple SDGs, such as SDG#1 no poverty, #2 zero hunger, #3 good health, well-being, etc. [4]. In general, current FS faces four critical issues [5]–[7]: production gap; triple burdens of malnutrition; inequity and inequality of FS outcomes; and Unsustainability of FS.

The first issue indicates that the current global agriculture and food system cannot produce enough food for the growing population. The second narrative has three sub-issues: over-nutrition or obesity, undernutrition, and micronutrient deficiencies [5]. The third issue implies that the current food system cannot create equal and equitable benefits for all, reflecting the underlying fact that 868 million people are starving. At the same time, one billion people are overweight globally, and the differences and imbalances between the global north and south regarding food security issues. The unsustainability of FS is the last narrative of FS failures, indicating one of FS's functions as a bridge that connects human beings and/or FS actors to the nature/biosphere, ecological and environmental systems, and the broad food environments. Moreover, Ericksen [8] argues that feedback loops exist between FS impacts and multiple global environmental changes (GEC). The outcomes and impacts of FS then promote or become triggers GECs. Meanwhile, GECs become risks or causes of FS issues.

Concerning those interactions and connections, it can be argued that FS issues and GECs can be solved simultaneously through a systematic and holistic approach. Applying the lens of circular economy strategies and food value chain (FVC) views will provide huge potential in solving urban FSs, promoting FS sustainability and urban sustainability, and eventually achieving circular urban metabolism. This study proposes a conceptual model of a circular urban food system (CUFS) to aid in implementing and promoting a circular economy in FS and cities. Modified 9R circular economy strategies are employed in the model respectively. They are refuse, rethink, reduce, reuse, refill, regift/ repurpose, recycling/recovery, rot, repeat. The objective of this study is to answer the following research questions:

1. How can CE principles be applied to urban FS through each stage of FVC?
2. How can such approaches address or reduce four narratives of FS failures?
3. How can these approaches be represented by a conceptual model (i.e., CUFS)?

To address these questions, this study conducts literature reviews on urban food systems/FS, CE principles and applications on FS in Section 2. The conceptual model and explanations are presented in Section 3. Followed by a discussion and conclusion in Section 4.

2 LITERATURE REVIEW

2.1 Urban food systems (UFS)

It is assumed that UFS has the utmost potential in FS transition towards sustainability to resolve most FS issues and reduce the impacts of climate change due to several reasons. First, cities are the leading contributor to climate change and are influenced by climate change the most at the same time, implying that the role of cities/urban areas in climate change actions is irreplaceable as an inescapable part of the solution to this GEC. Additionally, FS is also one of the main contributors to GHGs and are impacted by climate change greatly; indicating that FSs play an imperative role in developing climate change strategies. Lastly, since most of the world's population lives in cities, most of the world's food is also consumed in urban areas, tightly connecting the two critical actors of climate action. In short, it can be argued that FS of the cities or UFS holds the key to one of the most promising pathways for slowing or even reversing climate change impacts.

Speaking of UFS and FS definitions, there is no comprehensive, compact, and precise definition of UFS and no unified definition of FS in the literature. According to the Oxford Martin Programme on the Future of Food [12], FS can be considered a complex web of activities, including food production, processing, transport, and consumption. While the FS approach has been widely applied to identify, analyze, and assess the interactions, feedback, and impacts of FS actors and activities [12] and for examining the issue of GECs and food security, too [8]. On the other hand, Vieira et al. [13] define UFS as FS that encompasses activities and actors relating to cities' food supply and their interactions with the surrounding natural and built environments, socio-economic dynamics, and governance. With the growing recognition of the importance of FS in climate change and sustainable development around the world, other different types of FS have been introduced into academia and policymaking, such as sustainable food systems [14], [15], city region food system [16], [17], resilient food system [18], etc.

2.2 Circular economy principles and food system

CE principles originate from industrial ecology, which aims to close the loop of resource, material, and energy flows at the smallest possible cycles and decreases resource and energy inputs and waste outputs through reusing and recycling strategies [31]. In a recent article on CE and sustainability [33], several CE definitions are illustrated with a focus on different aspects of materials preservation and economic development. For instance, they define CE as "a regenerative system in which resource input and waste, emission, and energy leakage are minimized by slowing, closing, and narrowing material and energy loops, this can be achieved through long-lasting design, maintenance, repair, reuse, remanufacturing, refurbishing, and recycling" [34], highlighting the key features of "restorative and regenerative" [35]. While some authors defined a circular city as "is a city that practices CE principles to close resource loops, in partnership with the city's stakeholders (citizens, community, business and knowledge stakeholders), to realize its vision of a future proof city". In speaking of FS, a circular FS means it "prioritize regenerative production, favour reuse and sharing practices, reduce resource inputs and pollution and ensure resource recovery for future uses" [21].

According to ICLEI [21], there is an urgent need for our current dominant global FS to change due to the numerous environmental effects across the whole food value chain (FVC), including the following; 1. Clearing land and changing natural ecosystems for agricultural and food production remains the top cause of habitat and biodiversity loss; Agriculture occupies about 70% of global freshwater usage; Food and beverage occupy about 40% of all plastic packaging; Over 60% of the people's calories source come from only 4 crops (i.e., maize, rice, wheat, and potatoes), and monoculture of food reduce FS resilience dramatically.

Over of the 98% nutrients in food by-products and human waste generated in cities are directly landfilled, incinerated, or discarded without reusing or recycling;32% of all food loss and wastes happen in the stage of primary production; and 44% of all food loss and wastes occur in the stage of storage and processing, usually caused by insufficient cold chain.

ICLEI [21] argues that the food sector is recognized as a significant contributor to climate change and biodiversity loss, while these two issues can be resolved via CE solutions. When applying circular economy principles to each stage of the food value chain (FVC), they offer various opportunities to address FS issues. They enable each actor to identify the parts that they can participate in transforming the FS [21]. Furthermore, Jurgilevich et al. [31] also argue that CE, such as supporting local food chains with food wastes reduction, closing the nutrient loops, and developing policies for recovering and reusing essential raw materials, will offer solutions in moving toward a sustainable FS. By applying CE principles to the food system, natural systems will be supported by food while localizing food production and clearing food waste [21].

On the other hand, food is perceived as an interconnected urban infrastructure with the growing recognition of its central role in our survival in the urban context and well-being, sustainability, and prosperity for all [29]. However, in recent years, urban FS has been increasingly recognized as not sustainable, equitable, or resilient [28]. Regarding CE principles, it can be argued that pursuing a circular UFS promotes the overall FS sustainability and benefits circular urban metabolism. Moreover, one of four guiding principles from the FAO framework for the Urban Food Agenda is "resilience and sustainability by supporting the principles of a circular bioeconomy…and reshaping development pathways by taking account climate risks and vulnerabilities in urban food system planning" [32]; implying the vital role of utilizing circular principles as a powerful strategy in promoting urban FS transitions towards resilience, inclusiveness, and sustainability.

3 RESULTS AND ANALYSES: CIRCULAR URBAN FOOD SYSTEM MODEL

The following conceptual model (Fig. 1) is intended to be used to illustrate possible pathways of applying CE principles on each stage of FVC, to offer potential strategies to deal with issues of urban FS (i.e., four narratives) and improve its resilience, inclusiveness, greenness, circularity, and sustainability. This model is inspired by and builds upon various models and approaches, including Food System Framework [8], City Region Food System Assessment by FAO and RUAF [17], Ellen MacArthur Foundation's City Self-Assessment Tool [20], ICLEI's Circular Development [21], Food Environment Framework [23], Ellen MacArthur Foundation's Cities and the Circular Economy for Food [38]. The model consists of three belts (Section 3.1) and one core (Section 3.2). Detailed illustration and related key strategies are discussed in the following sections.

3.1 The triple belt of circular urban food system (CUFS)

According to FAO [39], a food value chain (FVC) "consists of all the stakeholders who participate in the coordinated production and value-adding activities that are needed to make food products". It has been divided into several stages regarding specific contexts and/or issues. For instance, in a conceptual framework for understanding the impacts of agriculture and food system policies on nutrition and health, Kanter et al. [40] evaluate FS through seven stages of FVC, including the origin of source, production, processing, distributing, storage, marketing, and consumption. While in another conceptual framework of food systems for children and adolescents, Raza et al. [41] employed disposing and wastes stage of FVC but no storage as follows: farm level, post-harvest, processing and packaging, transportation, trade, preparation, and disposing/wastes. Some only cover five stages of FVC when constructing a food supply system respectively, they are agricultural production; food storage, transportation, and trade; food processing and transformation; food retail and provisioning; and food consumption [30].

3.1.1 The belt of food value chain (FVC)

To demonstrate a complete picture and holistic view of the CUFS, the first belt is designed to cover the following seven stages of FVC (see Fig. 2).

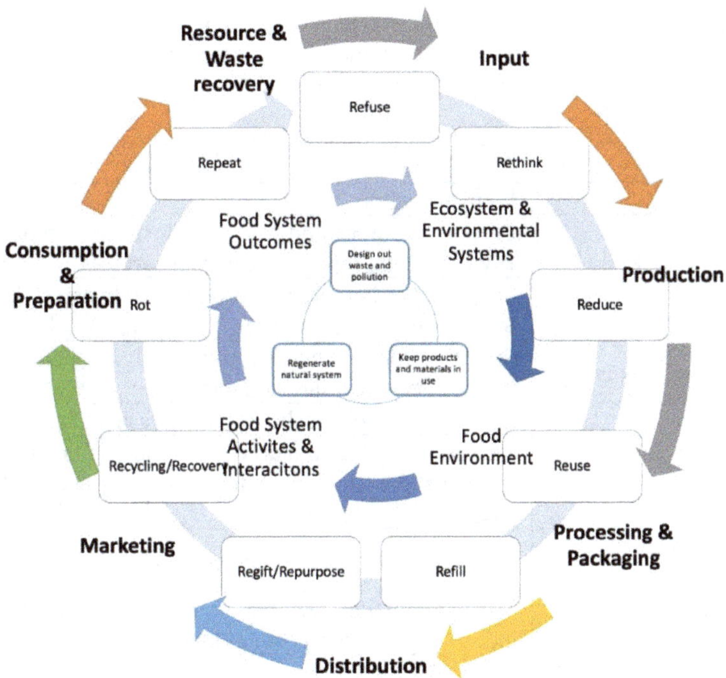

Figure 1: A conceptual model for a circular urban food system.

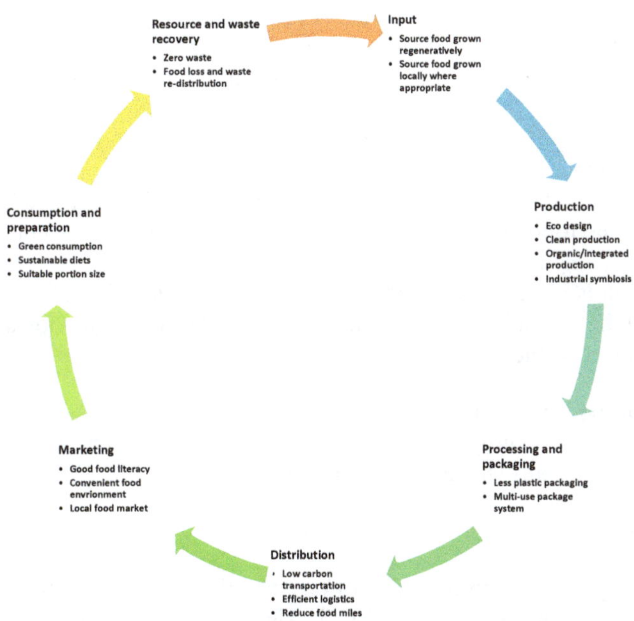

Figure 2: The belt of food value chain.

The second belt is to help identify the scope and coverage of the CUFS, it is suggested seven stages of FVC from cradle to cradle should be considered in constructing CE strategies regarding to the complex UFS issues (see Table 1).

Table 1: Seven stages of FVC.

	Elements/actors
Input	Land use, climate and weather, biosphere/ecological and environmental inputs, Ecosystem services
Production	Crop production, orchard production, livestock husbandry, forestry, managed fisheries (aquaculture)
Processing and packaging	Industrial food processing and packaging, home or warehouse processing and packaging
Distribution	Importers, exporters, brokers, wholesalers
Marketing	Retailers, local food markets, second-hand markets, sharing and exchange platform, food services and advertising, restaurants
Consumption and preparation	Catering, restaurants, home cooking
Resource and waste recovery	Source reduction, feed hungry people, feed animals, industrial uses, composting, landfill/incarnation

3.1.2 The belt of circularity

The CE consists of 12 different functions or activities. Circular Economy Asia [42] called them the "Circular Rs", including refuse, rethink, reduce, redesign, return, reuse, repair, refurbish, remanufacture, repurpose, renovate, recyclable resources, and recover. The Circular Rs are seen as standalone activities, even though they are already practiced in the linear economy [42]. Moreover, only very few can practice all the 12 Rs. For instance, in the CE business models illustrated by European Investment Bank [43], only 5Rs are considered: repair, reuse, refurbish, remanufacture, and recycle. While the Circulars City action framework proposed by ICLEI [21], utilizes a different set of 5Rs for its strategic directions, involving rethink, regenerate, reuse, reduce, and recover.

Since some of the 12 Rs are only suitable and essential to industry and business sectors, this study selects the following 9 Rs, which are the most applicable and practical for UFS to realize circularity along the whole FVC (see Fig. 3).

The second belt is to help identify the strategies for assisting CEUFS. It is suggested the nine Rs should be considered in promoting a circular economy towards sustainable transitions of FS (see Table 2).

3.1.3 The belt of food system interactions

The third belt is to help identify CE interventions of UFS at different scales. In detail, it has the following four elements (see Fig. 4 and Table 3).

3.2 The Core of CUFS

Since circularity is the main objective of our model, the core of CUFS, as illustrated in Fig. 5, is the three principles of CE [1], [38]: Design out waste and pollution; keep products and materials in use; and regenerate the natural system. For executing those three principles,

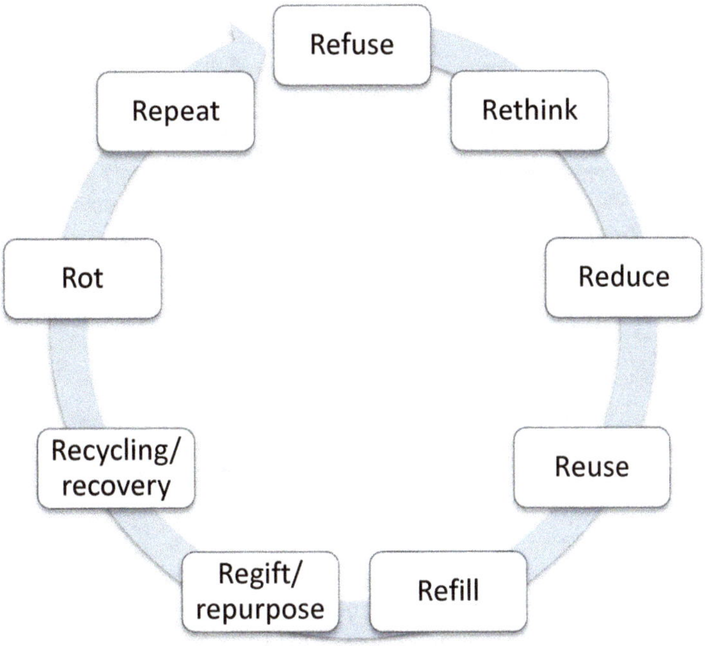

Figure 3: The belt of circularity.

Table 2: Nine Rs of CUFS.

	Descriptions/approaches
Refuse	Refuse to use unsustainable energy, materials, products, and process or there is a better alternative exists in terms of its greenness/eco-friendliness, eco-efficiency, footprints, and sustainability
Rethink	Redesign the UFS and FVC towards circular metabolism in a regenerative way
Reduce	Optimise the system and increase efficiency; reduce the consumption of energy, natural resources and material, food loss and wastes, resource and carbon footprint
Reuse	Design for extended use and reusable parts of products; improve people awareness
Refill	Design refillable products and packages; facilitate related infrastructures and regulation (e.g., water refilling station)
Regift	Facilitate second-hand markets and sharing and exchange platforms; Use a redundant product or its parts in a new product with a different function
Recycle	Recover materials from waste to be reprocessed into new things, improve the utility rate, promote upcycling, and minimize downcycling
Rot	Leverage microbial processes to turn food waste into municipal compost; biofertilizer and bioenergy
Repeat	Repeat the 9Rs cycle to promote CE strategies

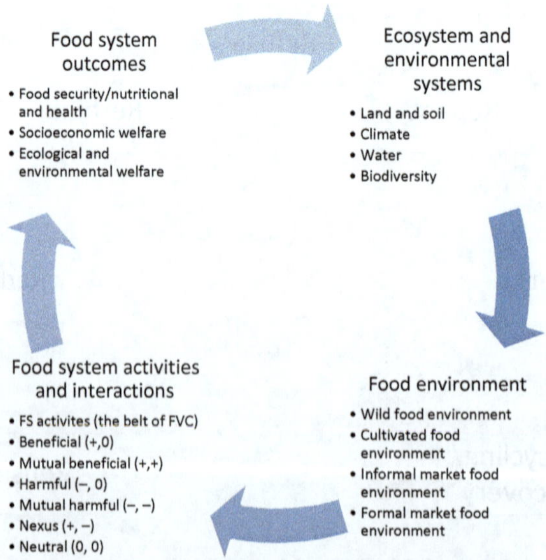

Figure 4: The belt of food system interactions. *(Source: Ericksen [8], Turner et al. [22], Ahmed et al. [23], and the authors, 2022.)*

Table 3: Four scales of FS interactions.

	Elements/actors
Ecosystem and environmental systems	Land use, climate and weather, biosphere/ecological and environmental inputs, ecosystem services (inputs of FVC)
Food Environment (FE)	Wild FE; cultivated FE; informal market FE; formal market FE
Food system activities and interactions	FVC activities (the belt of FVC); interactions: beneficial (+, 0), mutual-beneficial (+, +), harmful (–, 0), mutual-harmful (–,–), nexus (+,–), none (0, 0)
Food system outcomes/healthy and sustainable diets	Food security/nutritional and health; socioeconomic welfare; ecological and environmental welfare

the following seven themes are suggested to guide practical approaches and form a CUFS regarding the scopes and strategies from the belt of FVC (Section 3.1.1) and the belt of circularity (Section 3.1.2): eco-design; clean production; green consumption; zero waste management; equitable and equal benefits; inclusive and regenerative solution; and resilience capacity building.

It is argued that all these three core principles of CE are driven by design [1]. By considering pollution and waste are design defects [1], it's feasible to eliminate and/or avoid them at the beginning. Secondly, there are two main ways to keep products and materials in use: design the products reusable and refillable and use the materials in a circular loop.

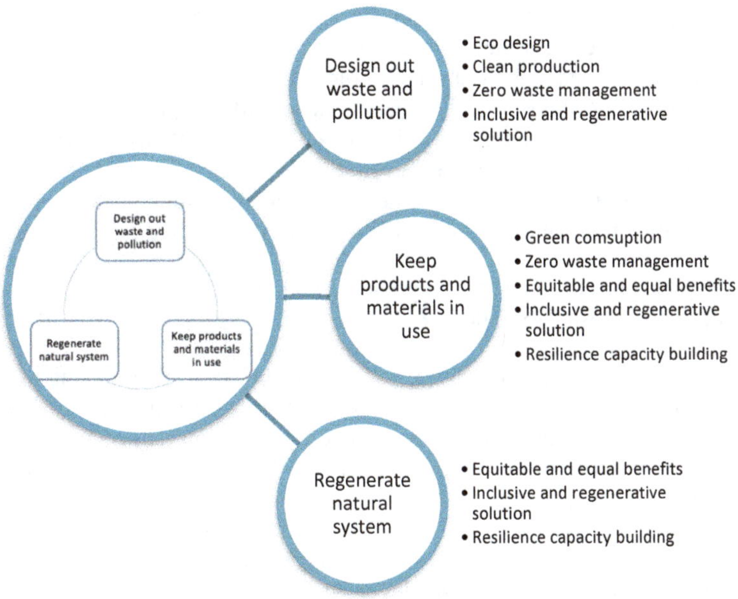

Figure 5: The core of circular urban food system. *(Source: Ellen Macarthur Foundation [1] and the authors, 2022.)*

Detailed strategies can be found in the belt of circularity (Section 3.1.2). Lastly, regenerating the natural system is the most crucial baseline of CUFS. ICLEI [21] interprets it as "harmonize with nature", highlighting the underlying basic rule of this principle is to ensure all infrastructure and interactions of producing and consumption stages benefits local resource and nutrient cycles while in deference to the natural ecosystems' regeneration rates.

4 DISCUSSION

This section discusses the results and considers to what extent the research questions have been answered in Sections 4.1–4.3.

4.1 Research Question: How can CE principles be applied to urban FS through each stage of FVC?

The three core principles of CE [1], [38] illustrated in Section 3.3 are the fundamental basis for the CE principles of Rs [21], [42], [43]. It can be argued that the Rs principles encompass the three cores while the three cores are also integral to the Rs principles. For better visualization and more comprehensive reflection, the 9Rs principles/strategies are used to show the application coverage along the FVC in Table 3. The applications of three cores can be found in Fig. 5 in Section 3.2.

Rethink, reduce, and repeat can be applied to every stage of FVC. Concerning the basis of three core CE principles, the main idea of rethinking is to redesign [1]. Thus, it can be applied by redesigning the products, UFS, incentives, and policies. All the material and energy can be reduced at each stage if using appropriate technologies and techniques for the reducing strategy. For instance, they reduce the carbon footprint by promoting renewable energy in food practices, minimizing non-recyclable packages, reducing food miles by localizing food sources, etc. Similarly, refuse, reuse, refill and regift can be applied to most FVC stages, except the waste/disposal stage, subject to the rot strategy. Detailed options can be found in Section 3.1.2, the Belt of Circularity. In a word, CE principles can be applied to each stage of FVC in UFS primarily by design with the aid of institutional and political support. Approaches built upon these seven themes are suggested: eco-design, clean production, green consumption, zero waste management, equitable and equal benefits, inclusive and regenerative solution, and resilience capacity building (see Fig. 5).

4.2 Research Question: How can such approaches address or reduce four narratives of FS failures?

Since CE approaches aim to keep material and sources in use while reducing extra waste, it will boost productivity and efficiency, hence reducing the production gap and FS unsustainability. Only a few approaches can reduce the triple burdens of malnutrition for the triple burdens of malnutrition. For instance, provide a better food environment and smaller portions, increase food literacy and labelling, promote food sharing platforms, etc. Such approaches can encourage people with higher food security levels to eat less. In contrast, re-distribute and/sharing excessive food with the hunger may reduce the number of people suffering from obesity and hunger at some level. Additionally, reducing monoculture of food sources, favour local food, and education for improving food literacy can reduce the incidences of micronutrients deficiencies. Moreover, a repurposing strategy like food sharing can also benefit from lowering the third FS narrative – inequity and inequality. Since part of their root cause is uneven distribution and shortages of resources and materials, CE approaches offer opportunities to address the causes of FS inequity and inequality.

Lastly, the unsustainability of FS directly links to the environmental impacts of the current linear FS and/or FVC. Besides improving the efficiency and utilization rate, CUFS is also dedicated to respecting local natural ecosystems' regeneration rate [21], contributing to a long-lasting balance between FS activities and natural environments. Moreover, CUFS' themes like eco-design, clean production, green consumption, zero waste management, and resilience capacity building can contribute to environmental sustainability directly. The other two themes, including equitable and equal benefits and inclusive and regenerative solutions, can promote social and institutional sustainability. And undoubtedly, CUFS has absolute advantages in terms of economic sustainability.

4.3 Research Question: How can these approaches be represented by a conceptual model (i.e., CUFS)?

CUFS consists of 3 belts and one core part to represent this conceptual model's scopes, strategies, scales, and subjects. The Belt of FVC (Section 3.1.1 and Fig. 2) indicates seven detailed stages covering the whole lifecycle of any food and/or food products to determine the scope of different contexts of issues. The Belt of Circularity (Section 3.1.2 and Fig. 3) represents 9 CE strategies that can be applied to multiple stages of FVC. The third Belt of Food System Interactions (Section 3.1.3 and Fig. 4) is designed to help define specific scales for different scenarios and situations, involving four main parts from the wider outside biosphere to abstract outcomes or individual activities. The core of CUFS (Section 3.2 and Fig. 5) is based on three core CE principles, aiming at providing guides to set specific subjects to each design of CUFS with seven proposed themes.

5 CONCLUSION

Current research on UFS with CE is minimal. The urgent need to resolve GECs, such as mitigating climate change impacts, food system issues, and rapid urbanization, enhancing the resilience, circularity, inclusiveness, and sustainability of the urban food system, brings a silver lining to address some of these GECs. With three core principles and 9 Rs strategies of CE, this study illustrates a conceptual model of CUFS. It provides potential pathways for developing such a system and offers optimal options for implementing CE interventions to boost the overall urban metabolism and urban sustainability. This study is limited by time, data scarcity, lack of literature, and case studies. Future directions may include indicator development for this conceptual model, case studies, circular cities, and CUFS, examining technological barriers of CUFS, roles of agroecology and/or urban agriculture in developing CUFS, trade-offs between circularity and efficiency, and so on.

ACKNOWLEDGEMENTS

The authors declare that they have no known competing financial interests or personal relationships that could have appeared to influence the work reported in this paper. This work has been kindly supported by the Green Chemicals and Technologies Scholarship of the University of Nottingham Ningbo China (reference number: 2003GCT_AC). The authors also thank the Ministry of Education, Culture, Sports, Science and Technology (MEXT), Japan for the provision of funding.

REFERENCES

[1] Ellen Macarthur Foundation, Circular economy introduction. https://ellenmacarthurfoundation.org/topics/circular-economy-introduction/overview. Accessed on: 21 Mar. 2022.

[2] ICLEI, Circular city actions framework. https://circulars.iclei.org/action-framework/. Accessed on: 21 Mar. 2022.

[3] Stefanovic, L., Freytag-Leyer, B. & Kahl, J., Food system outcomes: An overview and the contribution to food systems transformation. *Frontiers in Sustainable Food Systems,* vol. 4, Frontiers Media S.A., 2020. DOI: 10.3389/fsufs.2020.546167.

[4] United Nations, Food systems summit x SDGs. https://www.un.org/en/food-systems-summit/sdgs. Accessed on: 21 Mar. 2022.

[5] Allen, T., Prosperi, P., Cogill, B., Padilla, M. & Peri, I., A Delphi approach to develop sustainable food system metrics. *Social Indicators Research*, 141, pp. 1307–1339. DOI: 10.1007/s11205-018-1865-8. 2019.

[6] Bene, C., Oosterveer, P., Lamotte, L., Brouwer, I.D., de Haan, S., Prager, S.D., Talsma, E.F. & Khoury, C.K., When food systems meet sustainability: Current narratives and implications for actions. *World Development*, **113**, pp. 116–130, 2019.

[7] Li, X. & Siddique, K.H.M., Future smart food: Rediscovering hidden treasures of neglected and underutilized species for zero hunger in Asia. Food and Agriculture Organization of the United Nations: Bangkok, p. 242. 2018.

[8] Ericksen, P.J., Conceptualizing food systems for global environmental change research. *Global Environmental Change: Human and Policy Dimensions*, **18**, pp. 234–245, 2018. DOI: 10.1016/j.gloenvcha.2007.09.002.

[9] Verger, E.O. et al., A "fork-to-farm" multi-scale approach to promote sustainable food systems for nutrition and health: A perspective for the Mediterranean region. *Frontiers in Nutrition*, **5**, 2018. DOI: 10.3389/fnut.2018.00030.

[10] King, C. & Jaafar, H., Rapid assessment of the water–energy–food–climate nexus in six selected basins of north Africa and west Asia undergoing transitions and scarcity threats. *International Journal of Water Resources Development*, **31**(3), pp. 343–359, 2015.

[11] Laspidou, C.S., Mellios, N. & Kofinas, D., Towards ranking the water–energy–food–land use–climate nexus interlinkages for building a nexus conceptual model with a heuristic algorithm. *Water*, **11**(2), p. 306, 2019.

[12] Oxford Martin School, Oxford Martin Programme on the Future of Food. https://www.oxfordmartin.ox.ac.uk/food/. Accessed on: 23 Mar. 2022.

[13] Vieira, L.C., Serrao-Neumann, S., Howes, M. & Mackey, B., Unpacking components of sustainable and resilient urban food systems. *Journal of Cleaner Production*, **200**, pp. 318–330, 2018.

[14] Guarnaccia, P., Zingale, S., Scuderi, A., Gori, E., Santiglia, V. & Timpanaro, G., Proposal of a bioregional strategic framework for a sustainable food system in Sicily. *Agronomy*, **10**(10), p. 1546, 2020.

[15] El Bilali, H., Strassner, C. & Ben Hassen, T., Sustainable agri-food systems: Environment, economy, society, and policy. *Sustainability*, **13**(11), p. 6260, 2021.

[16] Blay-Palmer, A., Santini, G., Dubbeling, M., Renting, H., Taguchi, M. & Giordano, T., Validating the city region food system approach: Enacting inclusive, transformational city region food systems. *Sustainability*, **10**(5), p. 1680, 2018.

[17] Resource Centre on Urban Agriculture and Food Security (RUAF), City region food systems. https://ruaf.org/focus-area/city-region-food-systems/#:~:text=A%20City%20Region%20Food%20Systems,by%20strengthening%20rural%2Durban%20linkages. Accessed on: 21 Mar. 2022.

[18] Tendall, D.M., Joerin, J., Kopainsky, B., Edwards, P., Shreck, A., Le, Q.B., Krütli, P., Grant, M. & Six, J., Food system resilience: Defining the concept. *Global Food Security*, **6**, pp. 17–23, 2015.

[19] Wanjiru, R. & Neven, D., The NADHALI approach for assessing and planning city-driven food systems. *Measuring Impact*, p. 10, 2018.

[20] Ellen MacArthur Foundation, Cities and circular economy for food. https://ellenmacarthurfoundation.org/cities-and-circular-economy-for-food. Accessed on: 23 Mar. 2022.

[21] ICLEI, City Practitioners Handbook Circular Food Systems. https://circulars.iclei.org/wp-content/uploads/2021/03/ICLEI-Circulars-City-Practitioners-Handbook-Food.pdf. Accessed on: 21 Mar. 2022.

[22] Turner, C., Kalamatianou, S., Drewnowski, A., Kulkarni, B., Kinra, S. & Kadiyala, S., Food environment research in low-and middle-income countries: A systematic scoping review. *Advances in Nutrition*, **11**(2), pp. 387–397, 2020.

[23] Ahmed, S., Downs, S.M., Yang, C., Chunlin, L., Ten Broek, N. & Ghosh-Jerath, S., Rapid tool based on a food environment typology framework for evaluating effects of the COVID-19 pandemic on food system resilience. *Food Security*, **12**(4), pp. 773–778. 2020.

[24] Johnston, J.L., Fanzo, J.C. & Cogill, B., Understanding sustainable diets: A descriptive analysis of the determinants and processes that influence diets and their impact on health, food security, and environmental sustainability. *Advances in Nutrition,* **5**(4), pp. 418–429, 2014.

[25] Chen, P.J. & Antonelli, M., Conceptual models of food choice: influential factors related to foods, individual differences, and society. *Foods*, **9**, 2020. DOI: 10.3390/foods9121898.

[26] Biehl, E., Buzogany, S., Baja, K. & Neff, R.A., Planning for a resilient urban food system: A case study from Baltimore City, Maryland. *J. Agricu. Food Syst. Commun. Dev.*, **8**, pp. 39–53, 2018. DOI: 10.5304/jafscd.2018.08B.008.

[27] Lambrou, A., Berry, I., Hecht, A. & Labrique, A., A global food systems framework for pandemic prevention, response, and recovery. *Journal of Agriculture, Food Systems, and Community Development*, **10**(2), pp. 1–6, 2021.

[28] Jensen, P.D. & Orfila, C., Mapping the production-consumption gap of an urban food system: An empirical case study of food security and resilience. *Food Security*, **13**(3), pp. 551–570, 2021. DOI: 10.1007/s12571-021-01142-2/.

[29] Fassio, F. & Minotti, B., Circular economy for food policy: The case of the RePoPP project in the city of Turin (Italy). *Sustainability*, **11**(21), 2019. DOI: 10.3390/su11216078.

[30] Stuiver, M. & O'Hara, S., Food connects Washington DC in 2050: A vision for urban food systems as the centerpieces of a circular economy. *Sustainability*, **13**(14), 2021. DOI: 10.3390/su13147821.

[31] Jurgilevich, A., Birge, T., Kentala-Lehtonen, J., Korhonen-Kurki, K., Pietikäinen, J., Saikku, L. & Schösler, H., Transition towards circular economy in the food system. *Sustainability*, **8**(1), p. 69, 2016.

[32] Food and Agriculture Organization of United Nations, FAO framework for the urban food agenda (in brief). http://www.fao.org/3/CA3143EN/ca3143en.pdf. Accessed on: 21 Mar. 2022.

[33] Nikolaou, I.E., Jones, N. & Stefanakis, A., Circular economy and sustainability: The past, the present and the future directions. *Circular Economy and Sustainability*, **1**(1), pp. 1–20, 2021.

[34] Geissdoerfer, M., Savaget, P., Bocken, N.M. & Hultink, E.J., The circular economy: A new sustainability paradigm? *J. Clean. Prod.,* **143**, pp. 757–768, 2017.

[35] Ellen MacArthur Foundation, Circularity indicators. https://www. ellenmacarthurfoundation.org/programmes/insight/circularity-indicators. Accessed on: 3 Mar. 2021.

[36] Prendeville, S., Cherim, E. & Bocken, N., Circular cities: mapping six cities in transition. *Environ. Innov. Soc. Transit,* **26**, pp. 171–194, 2018.

[37] Ellen Macarthur Foundation, Cities and the circular economy for food. https://pacecircular.org/sites/default/files/2019-03/Cities-and-Circular-Economy-for-Food.pdf. Accessed on: 22 Sep. 2021.

[38] FAO, Sustainable food value chains knowledge platform. https://www.fao.org/sustainable-food-value-chains/what-is-it/en/. Accessed on: 24 Mar. 2022.

[39] Kanter, R., Walls, H.L., Tak, M., Roberts, F. & Waage, J., A conceptual framework for understanding the impacts of agriculture and food system policies on nutrition and health. *Food Security*, **7**(4), pp. 767–777, 2015. DOI: 10.1007/s12571-015-0473-6.

[40] Raza, A., Fox, E.L., Morris, S.S., Kupka, R., Timmer, A., Dalmiya, N. & Fanzo, J., Conceptual framework of food systems for children and adolescents. *Global Food Security-Agriculture Policy Economics and Environment*, **27**, 100436, 2020. DOI: 10.1016/j.gfs.2020.100436.

[41] Circular Economy Asia, The circular Rs. https://www.circulareconomyasia.org/circular-rs/. Accessed on: 24 Mar. 2021.

[42] European Investment Bank, The European Investment Bank circular economy guide supporting the circular transition. https://www.eib.org/attachments/thematic/circular_economy_guide_en.pdf. Accessed on: 21 Mar. 2022.

SECTION 2
ENERGY PERFORMANCE

OVERVIEW OF THE PERFORMANCE OF AN EXTERNAL VACUUM INSULATION COMPOSITE SYSTEM

NUNO SIMÕES[1,2], MÁRCIO GONÇALVES[2,3], CATARINA SERRA[2,3] & INÊS FLORES-COLEN[4]
[1]Department of Civil Engineering, University of Coimbra, CERIS, Portugal
[2]Itecons, Portugal
[3]University of Coimbra, CERIS, Portugal
[4]Instituto Superior Técnico, CERIS, Portugal

ABSTRACT

The external thermal insulation composite system (ETICS) is a construction technology commonly used for insulating building walls. Incorporating a super-insulation material, such as vacuum insulation panel (VIP), ETICS is a solution with great potential for improving the thermal performance of façades towards achieving nearly-zero energy buildings. However, the application of VIPs in buildings presents several challenges that need to be taken into account. Namely, those associated with handling and installation issues, design factors, as well as with the edge thermal bridging that occurs between panels, the doubts surrounding long-term performance and the high investment costs of vacuum technology. Also, conventional ETICS often presents early-signs of anomalies that could be worsened by using VIPs. Thus, many aspects need to be evaluated before VIP based ETICS become a viable solution. The main goal of this paper was to study the feasibility of incorporating a VIP solution into ETICS. For this purpose, an extensive investigation into the solution was carried out, following mainly experimental approaches. First, focus was put on the VIP thermal performance, in particular regarding the edge thermal bridging effect. Then, VIP-based ETICS walls were evaluated in terms of the whole system hygrothermal performance and durability. Real onsite walls and laboratory large-scale test specimens were assessed. New experimental procedures were defined to evaluate the durability of the solution and to enable the early identification of potential anomalies, in particular when exposed to solar radiation. Lastly, the cost-effectiveness of vacuum technology was analysed by means of a life cycle costing assessment. This research indicates that VIPs can be successfully used in ETICS. Nevertheless, such integration needs to be meticulously performed, since concerns specific to VIP installation need to be taken into account.

Keywords: vacuum insulation panels, ETICS, hygrothermal performance, experimental testing, onsite monitoring, life cycle costing.

1 INTRODUCTION

The EU 2030 Climate Target Plan is now aiming at reducing greenhouse emissions in at least 55% (from 1990 levels) and improving energy efficiency in at least 32.5% [1]. These ambitious targets are leading to increase the thermal requirements for buildings envelope over the last years. As a consequence, there is a need to use high thicknesses of thermal insulation materials, especially in cold climates, to accomplish the requirements of nearly-zero energy buildings. However, a thick envelope may not be desirable for several reasons, including architectural/design limitations and reduction of available floor area. Hence, there is growing interest in developing products using materials that are able to present a high thermal performance at a reduced thickness. In this field, emphasis is given to vacuum insulation panels (VIPs). The VIP consists in an evacuated open core material surrounded by thin laminates, composed of a barrier envelope (multi-film layer), used to maintain the vacuum. Typical values of VIP thermal conductivity (measured at centre of the panel) can be less than 5 mW/(m·K) in initial state, and 7 to 8 mW/(m·K) after 25 years [2]. These values are much lower than that of conventional thermal insulation materials, such as expanded polystyrene (EPS) that usually have thermal conductivity values around 35 mW/(m·K).

WIT Transactions on The Built Environment, Vol 210, © 2022 WIT Press
www.witpress.com, ISSN 1743-3509 (on-line)
doi:10.2495/ARC220031

Using this super insulating material in solutions widely spread in buildings could be a key factor to improve the thermal performance of buildings walls. For example, combine the benefits of external thermal insulation composite systems (ETICS) with the outstanding thermal performance of VIPs could have a great potential. Nevertheless, the use of VIPs in buildings have specific challenges that needs to be taken into account. For example, the panels: cannot be cut or adapted on-site; are very sensitive to mechanical damage; raise doubts regarding their service life performance; and have a high cost, as previously identified by Gonçalves et al. [3].

The present work aimed to study the feasibility of incorporating a VIP solution into ETICS. For this purpose, experimental approaches were conducted in order to evaluate the hygrothermal behaviour of the solution and its durability. First, the thermal performance of the VIP product was investigated, with particular attention to the edge thermal bridging effects. Then, two real VIP-based ETICS were built in different climates, namely in Warsaw (Poland) and Coimbra (Portugal). These case studies were monitored during two years, looking to evaluate long-term performance and detect early anomalies such as cracking, vacuum loss, blistering and biological growth. In laboratorial environment, the hygrothermal performance of the innovative solution was carried out for evaluating its durability and for enabling early identification of potential anomalies. In this context, a new test procedure that includes multiple dynamic boundary conditions, such as air temperature, relative humidity, rain and solar radiation simulation, was proposed and carried out on a large-scale test specimen. Additionally, since the high cost of vacuum technology is one of the most issues of it using in buildings, a life cycle costing analysis of the VIP is presented. Finally, some final remarks are taken.

During this research work, the VIP performance was often compared with a conventional thermal insulation material in order to evaluate the competitiveness of this innovative solution against the most commonly used ETICS solution, namely EPS based ETICS. Given the extension of the research, this paper only presents some of the main results achieved in this investigation.

2 MATERIALS AND METHODS

2.1 Description of the solution

The product used in this study consists in a vacuum insulation panel composed by fumed silica powder encapsulated in a metallized high barrier film. The nominal thickness of the panel is 20 mm. To allow for an ETICS application, the VIP product is fully encapsulated with a protective cover layer made of graphite expanded polystyrene with a thickness of 10 mm on each side and 20 mm along the edges. The EPS cover layer allows for the adhesive bonding of the panels onto substrate and provides additional mechanical resistance, minimizing the risk of perforation of the VIPs. The 20 mm of EPS at the edges of the panels also allow for the use of auxiliary mechanical fixing (plastic anchors), as can be observed in Fig. 1, and to allow for small size adjustments. The remain ETICS components (adhesives, base coat mortar, fibre-glass mesh and finishing coat products) are commercial products commonly used in EPS based ETICS solutions.

2.2 Thermal performance

Thermal transmittance measurements were performed on a guarded hot plate apparatus (GHP), λ-Meter EP500e, following ISO 8302 [4] and EN 12667 [5]. The tests were carried

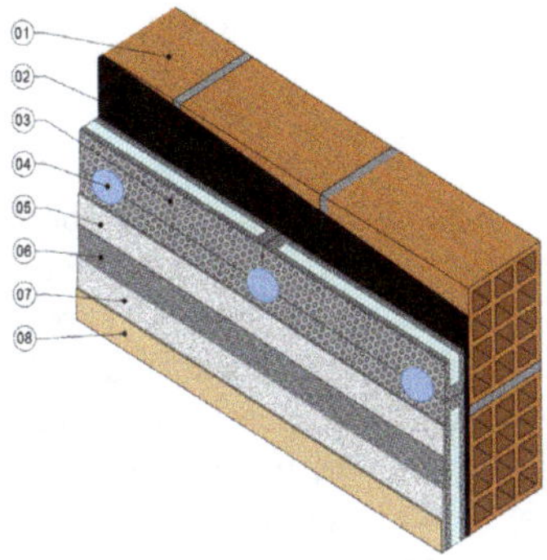

1. Support wall
2. Adhesive
3. Encapsulated VIP
4. Plastic anchor
5. Base coat (first layer)
6. Reinforcement (glass fibre mesh)
7. Base coat (second layer)
8. Finishing coat (primer and decorative coating)

Figure 1: VIP based ETICS scheme.

out in a controlled laboratory environment with (23 ± 2)°C air temperature and (50 ± 5)% air relative humidity. First, the thermal conductivity at the centre of panel (λ_{CoP}) was determined for each specimen. Next, the linear thermal transmittance between panel joints was measured. For this purpose, two VIPs were assembled within the GHP apparatus so that their joint was within the measuring area. The linear thermal transmittance, ψ, was determined by using the equation C.20 of EN 17140 [6]. Next, the equivalent thermal conductivity of the panel was determined according to equation C.23 of EN 17140 [6].

Besides the edge thermal bridging effects, the point thermal transmittances caused by mechanical fixation also need to be included in the effective thermal transmittance. For this reason, the methods described in EOTA TR 025 [7] and ISO 10211 [8] were followed and TRISCO software version v14.0w by Physibel [9] was used to determine the point thermal transmittance.

Then, the effective thermal transmittance of VIP based ETICS walls, including linear and point thermal transmittance was calculated and compared with a conventional ETICS solution. More details about the methodology used to evaluate the edge thermal bridging effect of VIPs can be found in Gonçalves et al. [10].

2.3 Onsite monitoring

To evaluate the VIP based ETICS performance under service conditions, two case studies were built and monitored. The first case focused on the hygrothermal behaviour of VIP based ETICS. The second one, aimed to evaluate and compare the condensation risk of VIP based ETICS and to compare with other conventional ETICS solutions. Both cases were monitored for more than 24 months. Next, the case studies are briefly described.

2.3.1 Case study 1

This case study consists in an existing façade retrofitted with ETICS solution with VIP. The retrofitted building is the Herpetarium of the Warsaw Zoo, Poland. The façade was divided in two areas: a reference wall side, corresponding to a traditional ETICS with EPS, and a VIP based ETICS wall with 42 m^2. Both walls are adjacent and face NE. Temperature, humidity and heat flux sensors were applied in two areas of the external wall under renovation. Sensors were placed behind the insulation and over the insulation (behind the rendering layer). Additionally, two sensors are located inside the building to measure indoor air temperature and humidity, and one sensor was placed outside the building to measure outdoor air temperature and humidity. More details about this case study are provided in Gonçalves et al. [11].

2.3.2 Case study 2

In case study 2, two free-standing walls with 5 m × 2 m (width × height) were built in Coimbra, Portugal. Each wall faces north and south. The walls are sectioned into areas that include different thermal insulation materials with 8 cm thickness, namely, expanded polystyrene, thermal insulation mortar (TIM), expanded cork (ICB) and encapsulated VIPs with 4 cm thickness. Two different finishing colours (white and black) were used for the finishing coat. Each area was instrumented with thermocouples that register temperature over time. Sensors were placed along the following interfaces: behind the insulation; over the insulation; and at the wall's external surface layer. Also, a weather station was placed near to the walls. Additional details about this case study are given in Gonçalves et al. [12].

2.4 Hygrothermal cycles resistance

With the goal of evaluating the hygrothermal performance of the VIP based ETICS, as well as evaluating its durability and identifying potential anomalies, accelerated ageing tests were carried out under controlled conditions. For this purpose, a real scale test-specimen with 2.8 m × 2.8 m was built. The hygrothermal cycles resistance was carried out according to section 2.2.6 of EAD 040083-00-0404 [15]. Furthermore, a new test procedure was proposed and implemented, that includes solar radiation simulation. This test method, further detailed in Simões et al. [16], allows for evaluate the impact of the optical properties of the rendering system on the ETICS performance. For this reason, two finishing coat colours were considered, namely black and white.

2.5 Life cycle costing

Since the high cost of VIPs is one of the main obstacles to the widely use of these materials in buildings, this research also focuses on their economic viability. The main goal was to identify the conditions under which the panels become cost-effective when compared with conventional ETICS. The study uses a comparative methodology based on the European cost-optimal methodology framework published in the Delegated Regulation no. 244/2012 [13], [14], with the added benefits of the rental income taken into account, as made possible by whole-life costing approach of ISO 15686-5 [15].

This study looks to quantify the economic benefit the additional space savings, due to their thinner nature of VIP when compared with EPS (considering the same thermal resistance). Thus, considering an office full-service leasing perspective, this investigation included the additional rental income expected due to the space savings resulting from the lower thickness of the VIP installation. The analysis was carried out with climatic and economic data from the city of Berlin, because in Germany the vacuum technology is most developed and whose climate is favourable to the use of high levels of thermal insulation. Further details are given in Simões et al. [16].

3 RESULTS

3.1 Thermal performance

The average VIP thermal conductivity measured at 10°C at the centre of panels is 4.20 mW/(m·K). While the average thermal conductivity measured at 25°C and 40°C, increased 4.40 mW/(m·K) and 4.67 mW/(m·K), respectively. Additional measurements were performed on intentionally perforated panels (panels without vacuum), resulting in a thermal conductivity of 21.6 mW/(m·K), which was still lower than conventional insulation materials.

Regarding the encapsulated VIP (panel encapsulated in EPS), the average thermal conductivity of centre of panels is 7.5 mW/(m·K). The linear thermal transmittance resulting from the joints between two encapsulated panels is 19.77 mW/(m·K), 20.95 mW/(m·K) and 22.25 mW/(m·K), considering the temperatures at 10°C, 25°C and 40°C, respectively.

In order to evaluate the influence of the panel's size, the increment of the equivalent thermal conductivity of the VIPs due to the inclusion of edge thermal bridging effects is presented in Table 1 (as a function of the size of the panel). These values are based on the assumption that the vertical rims of the panels have seamless joints (ψ_{10} = 9.89 mW/(m·K)), while the horizontal rims are assembled with overlapping foils (ψ_{10} = 11.20 W/(m·K)).

Table 1: Percentage increment of the equivalent thermal conductivity of encapsulated VIPs due to the edge effect.

		Horizontal rims						
	(mm)	**200**	**300**	**400**	**600**	**800**	**1000**	**1200**
Vertical rims	**200**	114%	96%	87%	78%	74%	71%	69%
	300	94%	76%	67%	58%	54%	51%	49%
	400	84%	66%	57%	48%	44%	41%	39%
	600	74%	56%	47%	38%	33%	31%	29%
	800	68%	51%	42%	33%	28%	26%	24%
	1000	65%	48%	39%	30%	25%	23%	21%
	1200	63%	46%	37%	28%	23%	21%	19%

As expected, using larger panels will reduce the influence of the edge thermal bridging. Also, the shape of the panels is relevant as the ratio between CoP area and VIP joints length may change.

Table 2 shows the U-value of a ceramic brick wall with VIP based ETICS considering the thermal conductivity of VIP at CoP (including 25 year ageing) with and without the edge thermal bridging effects. Also, the effective U-value, that includes the linear and point thermal transmittance is presented. The point thermal transmittance influence was calculated considering a number of mechanical fixings for each panel of five per square meter for smaller panels and three per square meter for larger panels. Additionally, the values considering an EPS with 40 mm thickness (same as VIP solution) instead the VIP solution are presented in brackets for use as reference. Internal and external surface thermal resistances of 0.13 (m^2·K)/W and 0.04 (m^2·K)/W were considered.

3.2 Onsite monitoring

3.2.1 Case study 1

Temperature and heat flux measurements allowed to prove the higher thermal performance of VIP, comparing with EPS ETICS solution. Fig. 2 shows the heat fluxes measured behind the VIP insulation (at the CoP and the joint area) in the VIP wall, as well as in the EPS wall, for a representative week during cold period.

Table 2: Effective U-value of the VIP based ETICS walls and EPS ETICS wall (in brackets), expressed in W/(m²·K).

	Ceramic brick wall	
Panel size	440 mm × 440 mm	640 mm × 640 mm
U-value without edge thermal bridging effect	0.235 (0.568)	0.206 (0.568)
U-value with edge thermal bridging effect	0.286 (0.574)	0.244 (0.572)
Effective U-value	**0.311** (0.576)	**0.259** (0.573)

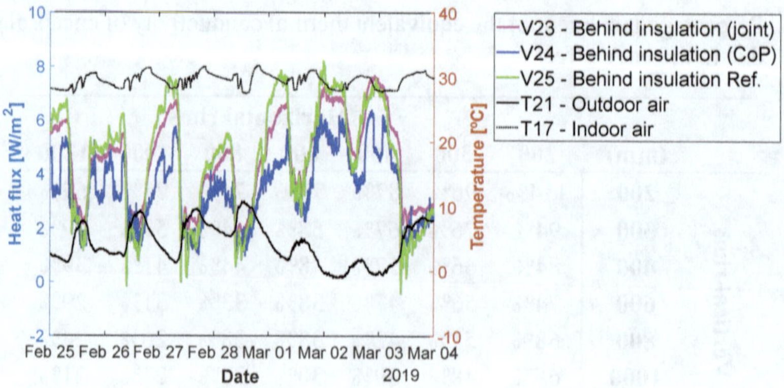

Figure 2: Heat fluxes and ambient temperatures curves during a winter period.

After 24 months of monitoring, regarding the appearance of ETICS rendering, no cracking, blistering, or aesthetic anomalies were found. Furthermore, infrared inspections were taken were taken periodically and from these thermograms, it is clear that the VIPs have not lost their vacuum during this period. In terms of humidity levels, the experimental measurements indicates that, when compared to the EPS wall, the VIP solution will not present additional problems regarding internal humidity levels.

3.2.2 Case study 2
Surface condensation issues contribute to microbiological growth, which is a common anomaly associated with ETICS. Condensation risk occurs when the superficial temperature is lower than the dew point. Thus, for one year, the time period in each month with surface temperature lower than the dew point temperature was determined. The annual condensation risk is presented in Fig. 3 for white and black walls (south orientation). The results suggest that walls with a higher level of insulation, that were finished with a white coloured surface are more likely to develop surface condensations that represents a higher biological growth risk in VIP ETICS walls.

Similar findings were found for the north-facing walls. However, in this case, the percentage of time in which there is risk of condensation increases around 18%, when compared with the south-facing walls.

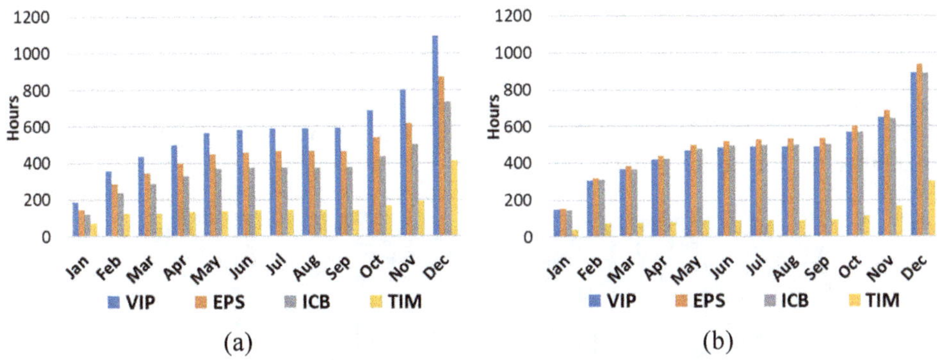

Figure 3: Accumulated hours in 2019 with surface condensation risk for different ETICS solutions. (a) White wall; and (b) Black wall.

3.3 Hygrothermal cycles resistance

Heat–rain and freeze–thaw standardized cycles were performed to assess the solution as resistant to hygrothermal cycle. After these hygrothermal cycles, no anomalies were found. Thus, according to the EAD for ETICS [17] this VIP based ETICS solution is assessed as resistant to hygrothermal cycles.

However, the new test that includes solar radiation cycles caused defects which were not found after the standard procedures, namely anomalies such as loss of flatness and finishing coat microcracking. This reveals the importance of studying the radiation effect on ETICS systems. Fig. 4 shows in detail the location of the microcracking on the finishing coat layer. Most of these cracks are located vertically and horizontally along the VIP joints. The cracking occurred only at within the solar radiation exposure area and its surrounding area.

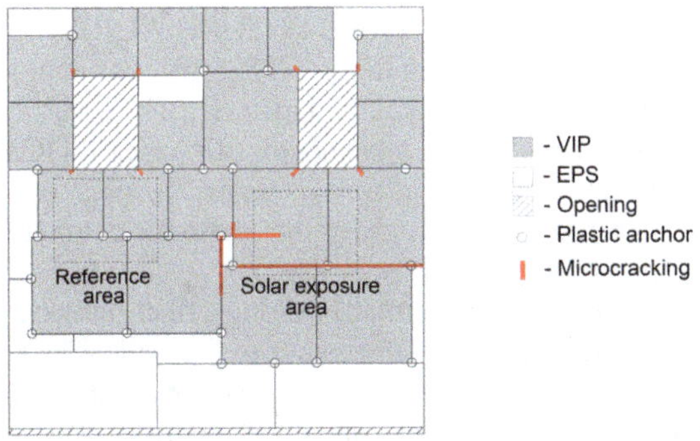

Figure 4: Layout pattern scheme of VIPs and location of the microcracking (in red) on the finishing coat layer.

Furthermore, during ageing cycles a loss of VIP thermal performance was verified. That is, the heat flux measurements carried out before and after the tests revealed a performance loss of about 70% compared to the initially estimated thermal transmission coefficient.

3.4 Life cycle costing

The cost-effectiveness of VIP solutions is strongly dependent on initial investment costs. Fig. 5 shows the results for the calculations performed considering an air conditioner system, a rental price area of 150, 250 and 350 €/(m^2.y) and a range in VIP prices between 1,500 €/m^3 and 3,000 €/m^3 (current price).

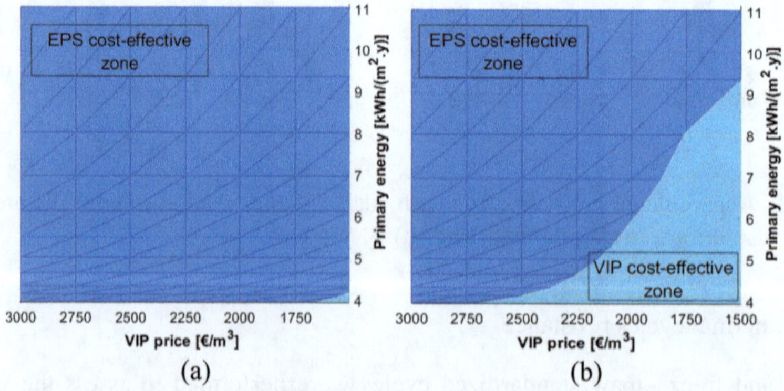

Figure 5: VIP price analysis for Berlin for financial perspective. (a) Fixed rental cost of 150 €/(m^2.y); and (b) Fixed rental cost of 250 €/(m^2.y).

Considering a low rental price area (150 €/(m^2.y) (Fig. 5(a))) it can be observed that the VIP is not profitable even when the VIP price is lower than 1,750 €/m^3. For a fixed value of rental cost of 250 €/(m^2.y) (Fig. 5(b)), if the cost of VIPs is reduced to less than 2,600 €/m^3, it may become a competitive solution against EPS, depending on the insulation level required.

4 CONCLUSIONS AND FUTURE WORKS

This paper presented a summary of results for an ETICS solution that incorporates vacuum insulation panels instead a conventional insulation material. The investigation focused on the study of the edge thermal bridges, on the analysis of VIP based ETICS performance under service conditions, on the implementation of accelerated ageing tests that simulates real conditions of exposure, as well as on the study of the economic competitiveness of the novel solution.

The results demonstrating that the thermal bridging effect between panel joints have a significant impact on the overall thermal performance of the walls. The installation (existence of air gaps), the size of the panel and the edge material design are determinant factors for reducing linear thermal transmittance values, and consequently, for promoting better overall thermal performance.

The higher thermal performance of VIP, compared to conventional insulations, was confirmed through temperatures and heat fluxes measurements, even when measuring at the joint area where the thermal bridging edge effect occurs. However, in comparison with other insulation materials, higher surface temperature amplitudes were found in the VIP solutions, increasing the risk of the rendering system cracking. Also, the condensation risk was found to be slightly higher when compared with an EPS ETICS solution, leading to higher biological growth risk on VIP based ETICS façades. After more than 24 months of onsite monitoring, the VIP based ETICS solution did not revealed remarkable anomalies. However, continue the long-term onsite monitoring campaign to evaluate the degradation of the thermal performance of the VIPs over time is suggested.

The proposed ageing cycles campaign revealed the importance of studying the solar radiation effect on ETICS systems. VIP joints could be considered the critical issue of the VIP based ETICS solution, since in addition to higher heat flux measurements verified at the joints, most of finishing coat microcracking occurred over joints. From the durability point of view, an increase of the VIP thermal conductivity was found. A deeper knowledge for the solar radiation impact on ETICS, even on innovative or conventional solutions, will be relevant to optimize solutions and assure a better long-term performance.

The use of VIPs in ETICS places greater demands for the rendering products, such as the need to resist to higher temperature amplitudes and slightly higher condensation risk, when compared to conventional ETICS solutions (as previously stated). So, there is a challenge to manufactures to develop rendering products able to mitigate these risks.

Replacing a conventional insulation by a vacuum panel allows a significant reduction in wall thickness and, consequently, an increase in the useful area. This space-saving economic benefit contributes to balance the high costs of the solution, from a perspective of long-term economic profitability analysis. This work shown the cost-effectiveness of VIPs, identifying the limits under which the panels become cost-effective against the use of expanded polystyrene.

Since the use of VIPs in ETICS are still few and recent, it is important to contribute to the experimental validation of this innovative solution, identifying the main limitations, promoting the optimization of products, and consequently increasing their competitiveness against current solutions. In this sense, this work contributes to increase knowledge in this field.

ACKNOWLEDGEMENTS

This work was supported by the Portuguese Foundation for Science and Technology (FCT), through the PhD grant PD/BD/135194/2017 under the Eco-Construction and Rehabilitation Programme.

REFERENCES

[1] European Commission, European Green Deal: Commission proposes transformation of EU economy and society to meet climate ambitions, 2021. https://ec.europa.eu/commission/presscorner/detail/en/IP_21_3541. Accessed on: 4 Dec. 2021.

[2] Jelle, B.P., Traditional, state-of-the-art and future thermal building insulation materials and solutions: Properties, requirements and possibilities. *Energy Build.*, **43**, pp. 2549–2563, 2011. DOI: 10.1016/j.enbuild.2011.05.015.

[3] Gonçalves, M., Simões, N., Serra, C. & Flores-Colen, I., A review of the challenges posed by the use of vacuum panels in external insulation finishing systems. *Appl. Energy*, **257**, 114028, 2020. DOI: 10.1016/j.apenergy.2019.114028.

[4] International Organization for Standardization, ISO 8302: Thermal insulation – Determination of steady-state thermal resistance and related properties – Guarded hot plate apparatus, 1991.

[5] European Committee for Standardization, EN 12667: Thermal performance of building materials and products – Determination of thermal resistance by means of guarded hot plate and heat flow meter methods – Products of high and medium thermal resistance, 2001.

[6] European Committee for Standardization, EN 17140: Thermal insulation products for buildings – Factory-made vacuum insulation panels (VIP) – Specification, 2020.

[7] European Organisation for Technical Approvals, TR 025: Point thermal transmittance of plastic anchors for ETICS, 2016.

[8] International Organization for Standardization, ISO 10211: Thermal bridges in building construction – Heat flows and surface temperatures – Detailed calculations, 2017.

[9] Physibel, TRISCO: 3D steady state heat transfer. http://www.physibel.be/v0n2tr.htm.

[10] Gonçalves, M., Simões, N., Serra, C., Flores-Colen, I., Rottenbacher, K. & Almeida, F.A., Study of the edge thermal bridging effect in vacuum insulation panels: Steady and unsteady-state approaches using numerical and experimental methods. *Energy Build.*, **258**, 111821, 2021. DOI: 10.1016/j.enbuild.2021.111821.

[11] Gonçalves, M., Serra, C., Simões, N., Flores-Colen, I., Kokolsky, C. & Sprengard, C., Onsite monitoring of a wall retrofitted with an external vacuum insulation composite system. *J. Build. Eng.*, **44**, 2021. DOI: 10.1016/j.jobe.2021.103301.

[12] Gonçalves, M., Simões, N., Serra, C., Almeida, J., Flores-Colen, I., Vieira de Castro, N. & Duarte, L., Onsite monitoring of ETICS comparing different exposure conditions and insulation materials. *J. Build. Eng.*, **42**, 103067, 2021. DOI: 10.1016/j.jobe.2021.103067.

[13] European Commission, Commission Delegated Regulation (EU) No 244/2012 of 16 January 2012 supplementing Directive 2010/31/EU of the European Parliament and of the Council on the energy performance of buildings by establishing a comparative methodology framework for calculating cost-optimal levels of minimum energy performance requirements for buildings and building elements. *Off. J. Eur. Union*, **28**, 2012. DOI: 10.3000/1977091X.C_2012.115.eng.

[14] European Commission, Guidelines accompanying Commission Delegated Regulation (EU) No 244/2012. *Off. J. Eur. Union*, pp. 1–28, 2012.

[15] International Organization for Standardization, ISO 15686-5: Buildings and constructed assets – Service life planning – Part 5: Life-cycle costing, 2017.

[16] Simões, N., Gonçalves, M., Serra, C. & Resalati, S., Can vacuum insulation panels be cost-effective when applied in building façades? *Build. Environ.*, **191**, 2021. DOI: 10.1016/j.buildenv.2021.107602.

[17] European Organisation for Technical Approvals, European Assessment Document: External Thermal Insulation Composite Systems with rendering, EAD 040083-00-0404, 2019.

BIOMIMICRY APPROACH DESIGN OF PETROL STATIONS WITH INTEGRATING RENEWABLE ENERGY IN THE UAE

ABEER ABU RAED[1,2], ALAA AHMED[1] & AMMAR ABDELHAMID[1]
[1]Department of Architecture, American University of Ras Al Khaimah, UAE
[2]Faculty of Architecture and the Built Environment, Department of Architectural Engineering and Technology, Delft University of Technology, The Netherlands

ABSTRACT

Highway petrol stations are located outside the main cities in the UAE. Connecting these remote petrol stations to power plants requires trenching for poles or underground cables as well as countless hours of work. The integration of PV solar panels and smart materials such as chromogenic glazing into the design and structure reduces the cost and the environmental impact. Even though solar energy is one of the cheapest sources of energy worldwide, it is challenging to achieve a design that reflects the identity of the city. Modern technology makes it difficult for designers and architects to balance themes such as sustainability, formation, and identity. The goal of the research is to explore the potential of using smart technologies to improve performance and to demonstrate, simultaneously, the most suitable design for optimizing energy for highway petrol stations in the UAE. As an example, the national Ghaf tree of the UAE is a cultural and historical symbol of stability and peace that can withstand harsh environment. This study introduces a reference design for the petrol station. A three-dimensional model using Autodesk Revit and an energy model using the Autodesk Insight program is built for the reference case. Several designs for the petrol station with different surface areas, shape, orientation, and window-to-wall ratio are studied to optimize the energy consumption. Variables such as the location of the oasis, its area, and its services will remain constant. The study concludes that the dynamic characteristics of smart glazing materials can combine not only one objective sustainable design quality, but more than one, including energy efficiency requirements to harmoniously integrate with the surrounding environment

Keywords: energy efficiency; optimized energy; building orientation, UAE.

1 INTRODUCTION

The harsh climate in the UAE causes high cooling demand throughout the year. Huge energy waste due to the use of low-efficiency appliances, high living standards, government subsidies, and the energy-intensive lifestyle have boosted the energy consumption in UAE [1]. A report by management consultancy Strategy &, formerly Booz Company, confirmed that electricity consumption alone over the past 10 years has more than doubled [2]. Therefore, energy-efficient buildings are a growing challenge, due to population growth, and high consumption rates that are unsustainable in the long term [3]. Proper design of the envelope components not only can save the required energy for the building but also can improve the thermal comfort of its occupants [4].

However, sustainable movement is growing to maintain a healthy lifestyle and overcome the rate of energy consumption that is growing rapidly in the UAE. In this context, Dubai Municipality issued buildings and glazing regulations supported by the law in 2003 and Abu Dhabi established ESTIDAMA Pearl building rating system in 2010. The building regulations implemented on newly constructed buildings will sustain the country with its rapid growth [5]. With the expo of 2020 UAE, there is a huge focus on introducing and maintaining the flow of the city with a sustainable approach by providing strategies of development to make UAE on the map of sustainability. In the cities of UAE, there had to be

WIT Transactions on The Built Environment, Vol 210, © 2022 WIT Press
www.witpress.com, ISSN 1743-3509 (on-line)
doi:10.2495/ARC220041

a modern approach to sustainability in the building sector which architecture forms. This will help in fulfilling the needs of the present without compromising the ability of future generations to meet their own needs [6].

Typically, a petrol station building includes production auxiliary rooms, service occupancy, office occupancy, and other facilities. The major source of energy consumption in petroleum stations is air-conditioning, lighting and electrical equipment. Petrol stations have a number of negative effects on the environment despite their importance. To reduce the environmental impacts of petrol stations, several researchers suggested sustainable designs of petrol stations [7]–[11]. Researchers in Europe, such as Pasi [7], Morales-Terrés et al. [8] and Ohlrogge and Wind [9], have focused on assessing the environmental impact of petrol stations rather than concerns regarding their energy use. Jovanović et al. [10] studied the topic of energy efficiency of petrol stations and the use of renewable energy for covering their energy demands. Abdullahi and Adedayo [11] investigated reducing the petrol station resources including raw materials, water, and energy for a petrol station in Nigeria.

This paper brings architecture as a new perspective to achieve sustainability in buildings and tests whether a building can be energy efficient by its architectural form. The paper presents a petrol station design as a future model for a highway petrol stations across the country. The importance of research lies in developing a design that reduces the energy consumption of the petrol station, which reduces material expenses and makes the station more sustainable through finding an ideal architectural form which contributes more to the environment and wastes less.

The work includes studying and analyzing of energy levels of a standard model. This will be followed by developing designs that improve the energy performance of highway petrol stations to reach the optimal form that reduces energy consumption in highways petrol stations across the UAE. The work aims at achieving a sustainable design that reflects the identity of the city.

2 METHODOLOGY

This paper compares three models; a standard model and two energy-efficient models (Design 1 and Design 2). Autodesk Revit will be used to generate the three-dimensional models. Autodesk insight will be used to study the energy performance of the three models considering different surface areas and shapes, orientation, surface area to volume ratio and glazing. Variables such as the location of the petrol station, its area, and its services will remain constant. A building's external envelope affects its thermal performance since it is directly exposed to the outdoor environment. Consequently, climate data including design external air temperature, average monthly external air temperature, wind speed and solar radiation will be fixed for all three scenarios. The distribution, the ratio and direction of the openings will change, which will be reflected on the energy performance of each design. Even cost and aesthetics are influenced by the building form. Undeniably, the selection of an optimum form, orientation, and envelope configuration can reduce energy consumption by 40% [12] because there is a high correlation between the shape of a building and its energy consumption [13].

3 STUDY AND ANALYSIS

The site is located in Dubai which is heading to the future by expo 2020. There is a movement in improving the technology and the infrastructure of the city to boost the economy and tourism. In addition, the urban aesthetic moving towards the future, through the metro station and the high-tech towers. The suggested location for the petrol station is in Ras Al Khor area which is accessed directly from Ras Al Khor Road. The site is surrounded by the industrial area from the south while the residential area is located to the east of the site. The site encounters the prevailing winds from the northeast with a speed of 14–18 km/h. The temperature can reach 41°C in summer, and it can drop to 18°C in winter.

The first design is driven by the futuristic image of the UAE. The aim is to create a new image or rebrand the petrol stations in Dubai. The design goal is to blend with urban aesthetic of Dubai, achieving a fast base metropolitan city. The assumption of a spaceship landing on sites of petrol stations is the fantasy that draws the shape of the petrol station Oasis. The colour choices are to make the oasis an eye-catcher on the highway and easy to notice also it pays tribute to the 1960s and 1970s futuristic aesthetic petrol stations in the USA (Norms, LA, California). The second design took a classical approach from the landscapes of UAE which is the desert. To achieve a variety in design proposals for the sake of the research a curved form had to be given. The dunes of the desert were the source of inspiration from which the first curves of the form were taken. A concept had to be added to the project to be familiar with the local taste in architecture with a contemporary look to serve as a new look for petrol station.

To study the energy performance of each design, a Revit model was generated. The energy simulation was conducted using Autodesk Insight with the assumption mentioned in Table 1. The three designs shown in Fig. 1 have the same floor area. The surface area of the reference design is 666.27 m^2 with area to volume ratio of 0.463 while the first design has surface area to volume of 0.492 and a surface area of 743.3 m^2. The second design has a 582.5 m^2 surface area with area to volume of 0.572 as presented in Table 2.

Table 1: Petrol station location and properties.

Location	Dubai, United Arab Emirates Latitude: 25.269510269165 Longitude: 55.3088417053223
Building envelop: Reference case	Exterior wall, U = 4.8794 W/(m^2 K) Roof, U = 0.602 W/(m^2 K)
Building envelop: Design 1 and Design 2 (exterior walls and roof)	U = 0.29 W/(m^2 K)
Glazing: Reference case	Single pane clear – no coating, U = 6.7069 W/(m^2 K)
Glazing: Design 1 and Design 2	Double low E, U = 1.26 W/(m^2 K)

Table 2: Petrol station design parameters.

	Floor area (m^2)	Surface area (m^2)	Volume (m^3)	Surface area to volume ratio
Reference case	360	666.27	1440	0.463
Design 1	360	743.3	1510.17	0.492
Design 2	360	582.505	1017.564	0.572

4 RESULTS AND DISCUSSION

The energy consumption for the three designs is presented in Fig. 2. The Reference design showed high energy consumption compared with the other two designs noting that the reference design does not follow the green building regulations in UAE while the alternative designs are implementing Dubai's green building code.

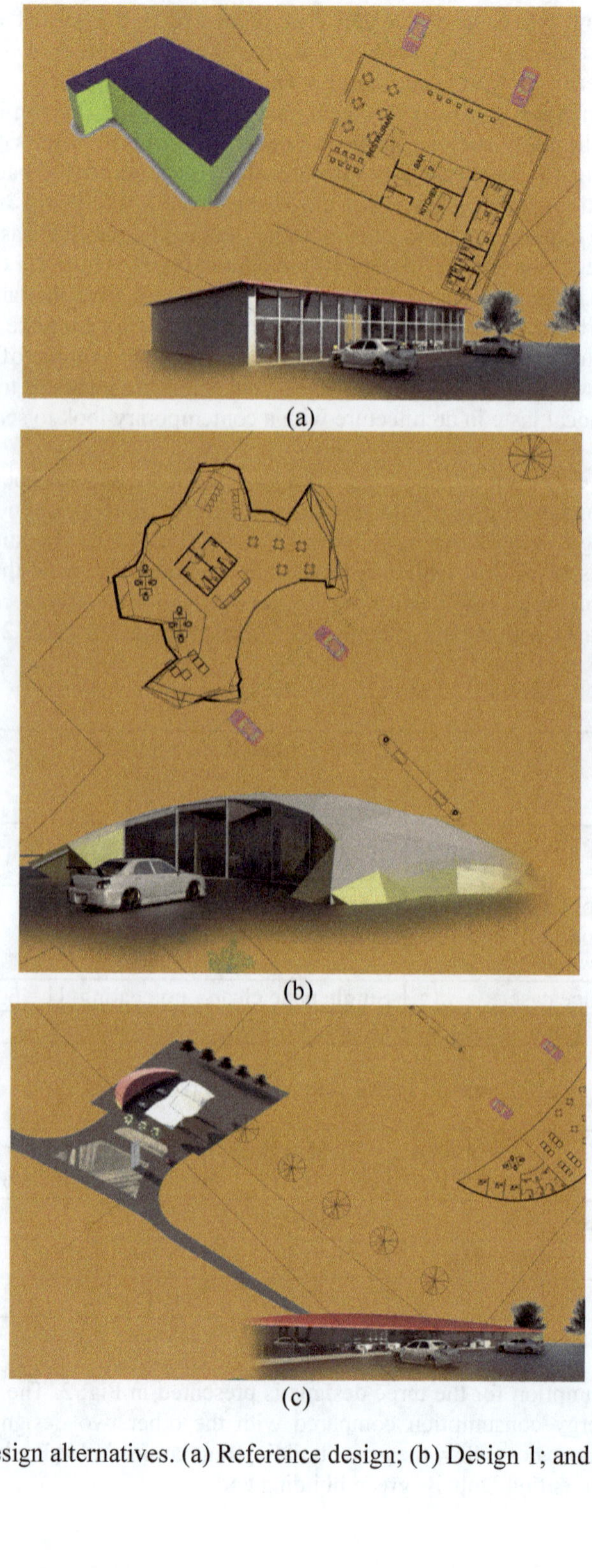

Figure 1: Design alternatives. (a) Reference design; (b) Design 1; and (c) Design 2.

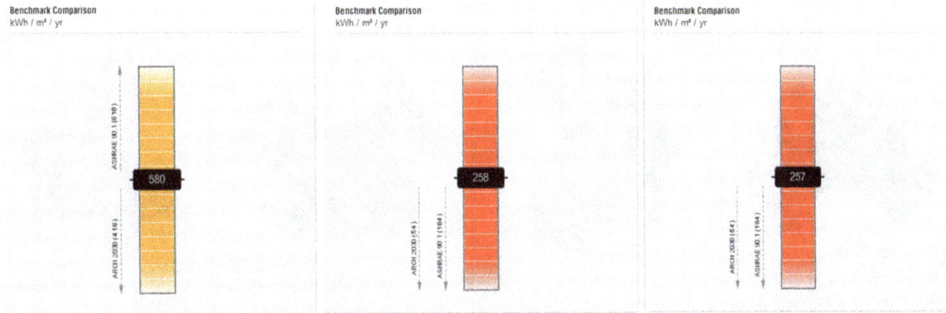

(a) (b) (c)

Figure 2: Energy consumption. (a) Reference design; (b) Design 1; and (c) Design 2.

Calculations indicate that the first and second design alternatives have the same energy consumption. The thermal gain is not the only consumer of energy. Factors such as thermal lighting and the proportions of openings in each direction play a big role in the rates of energy consumed. The energy consumption per year is 257 kWh per m^2 compared with 258 and 580 kWh per m^2 with Design 2 and the reference design respectively. The ratio of the surface area to volume of the first design is smaller than the ratio of Design 2. This reduces the heat transfer rate for the first design compared to Design 2 to minimize the thermal transfer across the building's exterior, the building should be as compact as possible, tilted toward the cubic shape. But to reach the optimum shape the three factors, temperature, wind speed and direction and the building orientation should be taken into consideration. To reduce the environmental impact of using traditional energy sources, solar panels will be integrated into the energy-efficient design. Nowadays, there is a wide range of developed and innovative solar systems that can be integrated into buildings and not just added to the roof. These products are varied in technologies, performance and form. Advanced products become not just additive elements, but part of the main envelope components.

It is important to consider the glazing system to reduce energy consumption and achieve a good indoor thermal environment. Heat transfer through windows is high because they are usually the weakest parts of the building membrane for heat transfer. Glass is considered a weak separator between the outdoor and indoor environment. The window-to-wall ratio has a clear effect on cooling energy. Windows orientation also has a major effect on heat transfer. In general, windows should occupy the minimum wall area in each direction in hot climates. In this context, Design 1 and Design 2 are further investigated to study the effect of orientation and glazing. As shown in Fig. 3, the effect of both glazing and orientation is minimal for Design 1. Only 3% difference in energy consumption between the best and the worst cases. Design 2 is more sensitive to glazing and orientation. The difference in energy consumption between the worst and the best case is in the range of 13%.

5 CONCLUSION

Petroleum stations play a key role in the development of the transportation sector. To achieve a green, low-carbon transportation system, petrol stations need to reduce their energy consumption. This study introduces a reference design for the highway petrol stations in UAE. The suggested designs integrate energy-efficient buildings with architectural concepts drawn from the classical approach of UAE which is the desert or the futuristic image of UAE.

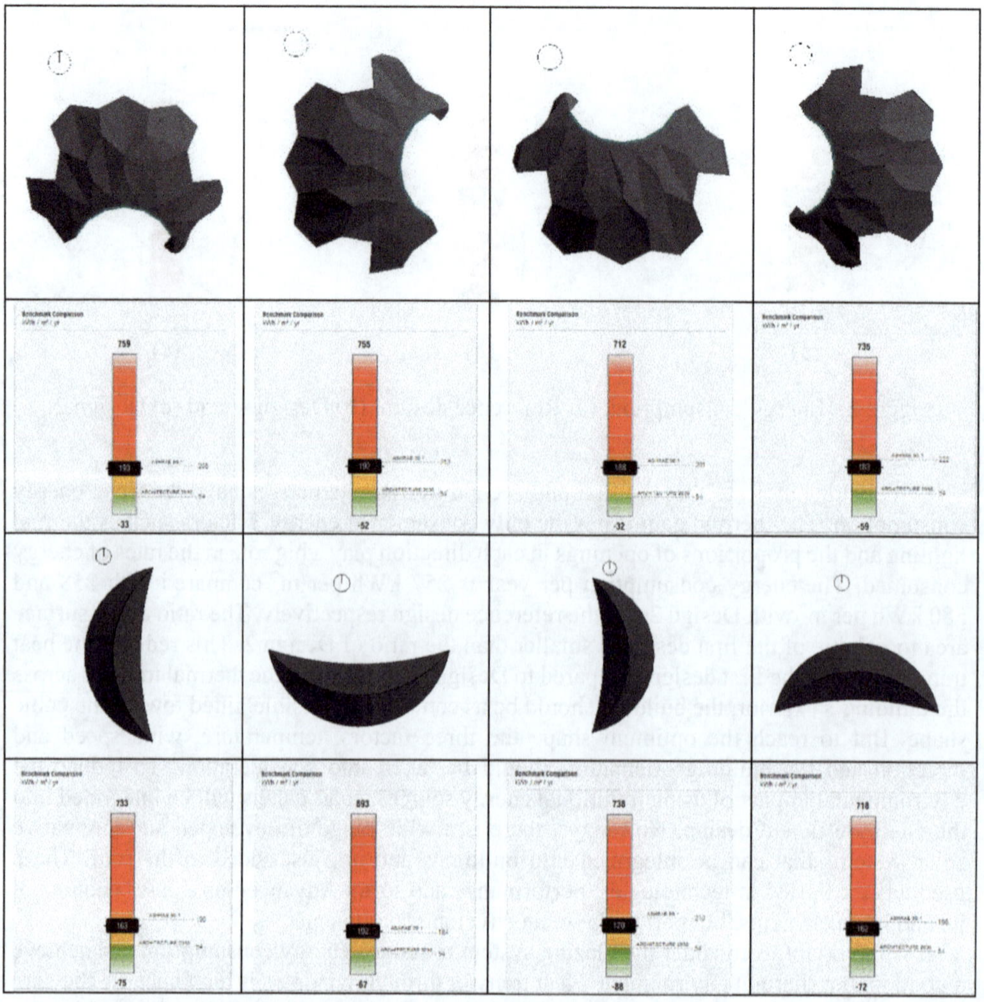

Figure 3: Effect of orientation and glazing.

The suggested designs were analyzed to study the reduction in energy consumption when implementing Dubai's green building code. The work revealed that adapting efficient energy measures in designing petrol stations in UAE can reduce energy consumption per m^2 from 580 kWh to 257 kWh per year. Furthermore, the effect of glazing and orientation on energy consumption was also studied. It was found a reduction of 13% in energy consumption can be achieved through orientation and glazing.

REFERENCES

[1] Sakhrieh, A., Energy analysis for GCC countries. *Global Conference on Renewables and Energy Efficiency for Desert Regions*, 4–6 April, Amman, Jordan, pp. 1–4, 2016.

[2] Maceda, C., UAE electricity consumption to grow "more rapidly". *Energy – Gulf News*, 29 Oct. 2018. gulfnews.com/business/energy/uae-electricity-consumption-to-grow-more-rapidly-1.1569200.

[3] Asif, M., Growth and sustainability trends in the buildings sector in the GCC region with particular reference to the KSA and UAE. *Renewable and Sustainable Energy Reviews*, **5**, pp. 1267–1273, 2016.

[4] Almarzouq, A. & Sakhrieh, A., Effects of glazing design and infiltration rate on energy consumption and thermal comfort in residential buildings. *Thermal Science*, **23**(5B), pp. 2951–2960, 2019.

[5] Friess, W. A. & Rakhshan, K., A review of passive envelope measures for improved building energy efficiency in the UAE. *Renewable and Sustainable Energy Reviews*, **72**, pp. 485–496, 2017. DOI: 10.1016/j.rser.2017.01.026.

[6] Steele, J., *Sustainable Architecture: Principles, Paradigms, and Case Studies*, McGraw-Hill: New York, 1997.

[7] Pasi, M.N., Environmental protection standards at petrol stations: A comparative study between Finland and selected European countries. Tampere University of Technology, Department of Civil Engineering, Engineering Geology, Tampere, 2005.

[8] Morales-Terrés, I.M., Doval, M.M., González, F.E., Baeza Caracena, A., Barberá, R.J., Assessing the impact of petrol stations on their immediate surroundings. *J. Environ. Manage.*, **91**, pp. 2754–2762, 2010.

[9] Ohlrogge, K. & Wind, J., New technology for emission reduction at petrol stations. *12th Regional Conference IUAPPA and 4th International Conference on Environmental Impact Assessment*, Praha, pp. 418–425, 2000.

[10] Jovanović, L., Fedjunina, D. & Radosavljević, M., The optimization of development process of petrol-gas stations network with the purpose to increase energy efficiency. *Zašt. Mater.*, **56**, pp. 100–104, 2015.

[11] Abdullahi, A.P. & Adedayo, O.F., Application of sustainable environmental design requirements in filling stations in Niger State, Nigeria. *Conflu. J. Environ. Stud.*, **11**(1), pp. 104–116, 2017.

[12] Muhaisen, A.S. & Abed, H.M., Effect of building proportions on the thermal performance in the Mediterranean climate of the Gaza Strip. *Journal of Engineering Research and Technology*, **2**, 2015.

[13] Ourghi, R., Al-Anzi, A. & Krarti, M., A simplified analysis method to predict the impact of shape on annual energy use for office buildings. *Energy Conversion and Management.*

IMPACT OF NANO INSULATING MATERIALS ON ENERGY RETROFIT OF BUILDINGS

HATEM F. SALEM, MOHAMED ABDELAAL IBRAHIM & ZEYAD EL-SAYAD
Department of Architecture, Faculty of Engineering, Alexandria University, Egypt

ABSTRACT

Climate change is one of the most significant challenges currently. Reducing greenhouse gas emissions and global energy demand has become an important research topic. These challenges have increased interest in retrofitting the existing building stock. Energy retrofitting, i.e., energy savings and optimal energy use, is necessary to close the gap between limited resources and increasing energy demand. Applying an energy-efficient insulation system can significantly decrease the energy consumed via a building's air-conditioning system during the summer. Hence, building insulation has become a promising research topic, especially insulation based on nanomaterials due to their low U-values. This research paper examines the energy retrofitting effectiveness of an old educational building in Egypt using Nanogel®Aerogel insulating material by integrating it with the building envelope, while walls are insulated with vacuum insulation panels (VIPs). The energy simulation was performed by DesignBuilder software (version 6.1.0.006). The results indicate that integrating VIP and aerogel in the building envelope can increase the thermal performance of a building in a hot climate like Egypt, which especially needs cooling loads during the summer months. It also shows a significant reduction in annual energy consumption, saving up to 36.5% compared to the base case.

Keywords: nano insulating materials, retrofit building, energy efficiency, simulation, aerogel, VIPs.

1 INTRODUCTION

Today, due to the rapid economic growth and high levels of living, the energy of fossil fuel procurement is significantly depleted, which has led to environmental pollution and an energy crisis. Existing buildings account for 40% of the total world's primary energy consumption and one-third of global carbon dioxide emissions [1]. This high energy consumption in buildings is mainly associated with the lower performance of the envelope of the building.

Retrofitting existing buildings offers a significant possibility for conserving energy and lessening emissions, because large numbers of the existing building stock have poor energy efficiency. Most of the existing buildings in Egypt were designed and constructed without specific energy performance requirements for building envelopes. Therefore, retrofitting existing buildings is one of the possible ways to improve energy efficiency. A well-designed envelope should use novel insulation materials to control the heat transfer and minimize it within the building. Using traditional insulating materials usually requires an increasing insulation layer on the wall, the floor and the roof. In retrofit projects, this, in turn, limit the ability to adjust the function and design and affects the ability to apply sustainable solutions depending on the required renewal level.

Accordingly, advanced solutions capable of combining energy efficiency with the quality of building architecture are needed, ensuring a higher comfort for its users. In that case, super insulation materials (SIMs) can achieve many benefits in existing buildings by improving their energy and environmental performance and increasing the building's life [2].

One type of SIMs is vacuum insulation panels (VIPs) and advanced nano-porous materials such as aerogel. VIPs have extraordinarily high thermal resistance values up to 20 times higher than conventional insulating materials (see Fig. 1); their significant advantage is that they can be available in thinner layers (approximately 5 to 50 mm thick) with low thermal conductivity values (0.015–0.004 W/mK) [1]

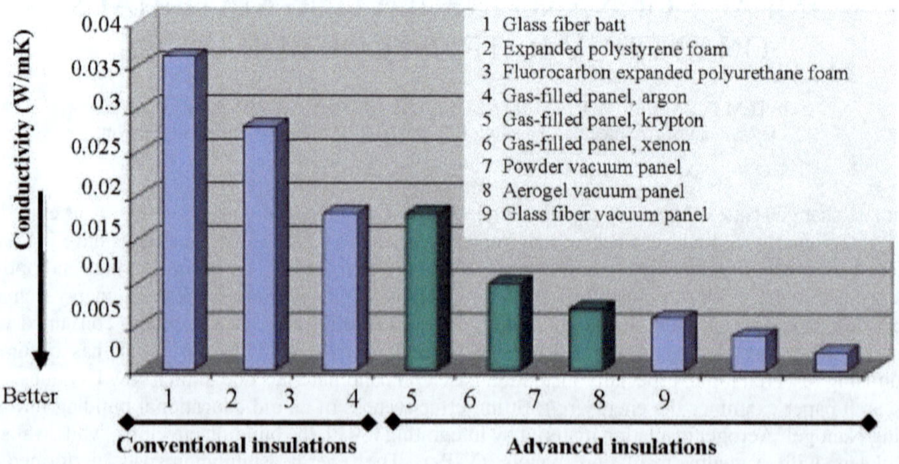

Figure 1: Thermal conductivities of conventional and advanced insulating materials [3].

VIP insulation consists of three main components: an inner porous core, a getter/desiccant, and an airtight envelope from plastic or aluminium, as shown in Fig. 2 [3].

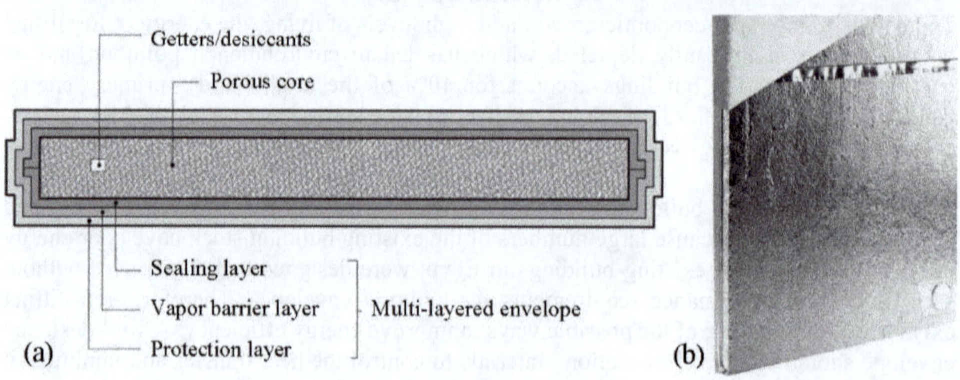

Figure 2: (a) Structure of a vacuum insulating panel; and (b) Vacuum insulating panel core and envelope [4].

The principle methodology that characterizes VIPs and allows for their exceptional performance is vacuum pressure, which can be used to decrease gaseous thermal conductivity that reduces heat transfer [4].

Aerogel is also applied in glazing as a granular aerogel encapsulated between panels; compared to the equivalent glass unit, it is 200 times higher in strength and 20% lighter in weight with low U-values (0.30 W/m^2K). Added to that, it has adequately optical transparency and increased acoustic insulation

Various research has been done recently to investigate the impact of nano insulation materials on building retrofits and evaluation of various energy retrofit strategies, including actual building studies, simulations, and laboratory experiments.

Ascione et al. [5] declared the dynamic testing of a concrete wall with VIPs in different weather conditions of the test building and furthermore conducted simulations of an office building model to assess the relative benefits of the very low conductivity of VIPs under multiple weather conditions. Challenges posed by using VIPs in external insulation systems were reviewed by Gonçalves et al. [3]. Another study in KSA by Mujeebu et al. [6] conducted a simulation study to determine the influence of applying aerogel (nanogel) glazing in windows and VIPs on external walls of a prototype office building on energy consumption; the study revealed that nanogel glazing alone could save 16% in building annual energy consumption.

Buratti and Moretti [7] studied silica granular nanogel in a glazing window with an aluminum frame prototype in an in-lab experiment. They discovered the extraordinary thermal performance of the material and its high acoustic properties.

In Egypt, sustainable design is still in its modest beginnings. A few numbers of architectural practitioners and building owners are interested in raising the energy efficiency of existing buildings stock and enhancing their internal comfort without consuming more harmful energy sources.

A current case study of retrofit of the building of the engineering faculty in Alexandria was selected to examine the impact of nano-insulating materials on the energy consumption of the building.

Firstly, a quick overview of the current reference case and location, followed by a methodology for the simulation stage and analysis of the different interventions used in the building compared to the base case by using DesignBuilder software for simulating the energy of each intervention, then the results and discussion, and finally, extract conclusions and recommendations to be a guideline for designers, researchers and buildings' owner.

2 THIS CASE STUDY: OVERVIEW

2.1 Faculty of Engineering, Alexandria University, Egypt

The case study building for this research is located in the Central district of Alexandria – El-Shatby – Egypt. The Engineering Faculty at Alexandria University (see Fig. 3) is operated as a typological representative for Alexandria's public institutions of architectural education.

The distinguished architect Dr. Mohamed Kamal Ismail designed the faculty's buildings and the studying initiated in the faculty in the 1942–1943 academic year. The building is old

Figure 3: The Engineering Faculty building at Alexandria University [8].

and deteriorated, while the preventative maintenance was inefficient. However, recently started, the restoration of the building's facades [8].

2.2 Alexandria climate

Egypt lies between latitudes 22° to 32° N and 25° to 35° E longitude. Egypt's climate is preponderantly hot. Egypt is divided by the Housing and Building Research Centre (HBRC) into eight various climatic design regions. According to Koeppen's climate classification, Alexandria is located in the north coast region and has a hot, humid climate. Alexandria's climate is hot in the summer, where the temperature can reach 32°C. or even 34°C, and the summer duration begins in May and ends in October. The winter period lasts from November to April [9].

3 METHODOLOGY

The current case study indicates how VIPs and Nanogel (Aerogel) can be used to retrofit old buildings and improve their energy efficiency performance. A simulation was applied using DesignBuilder software to determine the building's energy consumption and use intensity to achieve the most productive energy efficiency and investigate the effect of such nanomaterials on the building energy performance.

The presented study went through three primary simulation interventions. First, VIPs were added as exterior insulation to the building envelope; secondly, to achieve adequate energy performance, VIPs were used as interior insulation; and lastly, the old windows (3 mm single glass) were changed with double nano glazing windows. All the above-suggested interventions were simulated to compare results of the overall energy efficiency with the base case before retrofitting.

3.1 The base case

The southern side on the fourth floor of the building (see Fig. 4) was chosen for the current study for several considerations. It represents the primary floor of the architecture department which is diverse in functions (see Fig. 5); in addition, the zones range in cooling between air conditioning and natural ventilation; furthermore, part of this floor ceiling is an external roof. The southern facade was concerned with the study to demonstrate the results with more contrast and realism. Therefore, the current floor considers most factors affecting the energy performance of the building and represents a sample model of the rest of the buildings at the faculty.

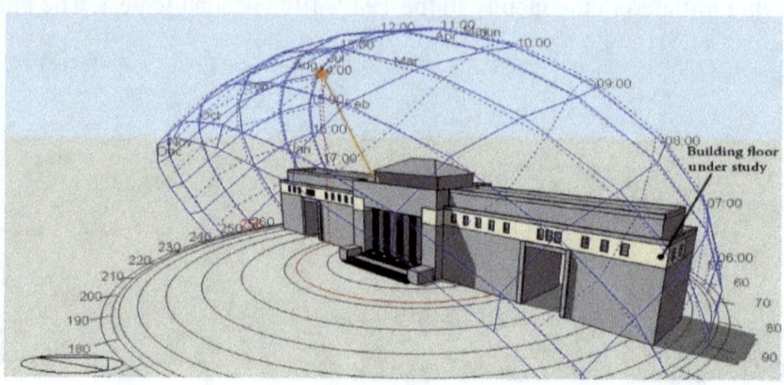

Figure 4: The base case simulation thermal model.

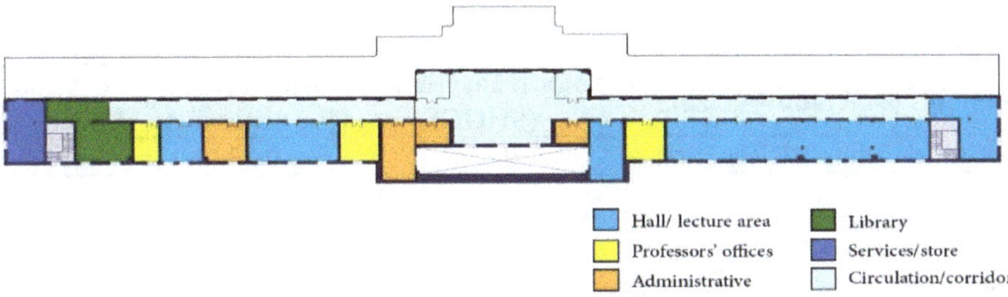

Figure 5: Southern side fourth level floor plan.

3.1.1 Constructions

The building has non-insulted brick masonry walls 500 mm thick covered with a 40 mm thick artificial sandstone on the exterior, providing a calculated U-value of 1.031 W/m^{2}K. The materials of the original building's envelope are described in Fig. 6.

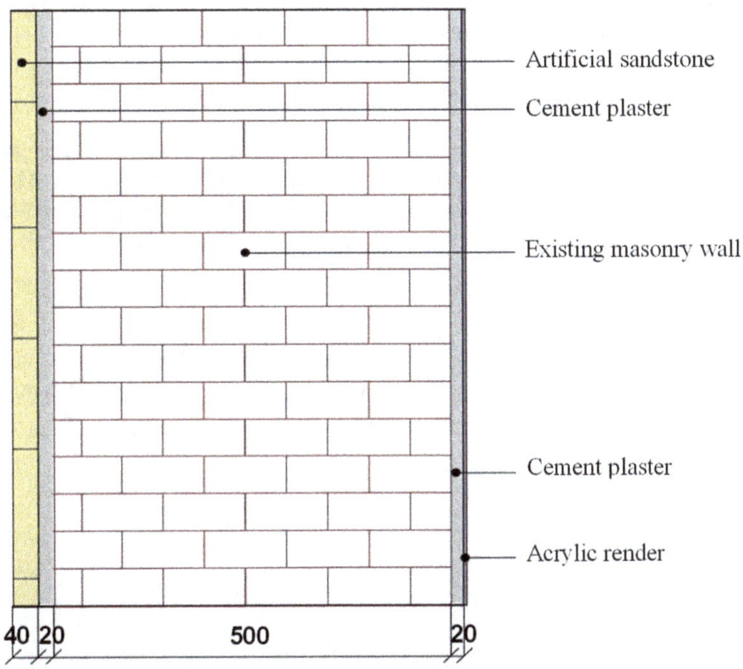

Base case external wall (non-insulted brick masonry walls 500 mm thick).
U-Value: 1.031 Wm^{2}K

Figure 6: The base case materials of external walls.

3.1.2 Exterior openings

The exterior windows of the building are wooden frames with a single clear 3 mm glazing. The original window has a U-value of around 6.2 W/m^{2}K (see Table 1).

Table 1: Base case windows characteristics. *(Source: DesignBuilder.)*

Symbol	Window glazing	Solar transmission (SHGC)	Direct solar transmission	U-Value (W/m^2-K)
G0	Single Clear 3 mm	0.858	0.837	6.257

3.2 Weather data

DesignBuilder software (version 6.1.0.006) was used as a model utilizing the dynamic thermal simulation for various walls, roofs and windows. Building's 2D AutoCAD file plan was imported to draw the building and conduct the simulation analysis. The simulation evaluated the cooling load's annual energy consumption (AEC); the space cooling system implemented in the case study is a (split no fresh air) in all the conditioned zones. The schedules were set using settings of fixed activities based on the traditional Egyptian lifestyle (holidays, occupancy times, work hours).

The climate data was set to Alexandria climate data. DesignBuilder software contains EPW weather files for Alexandria city. The EPW's weather file is extracted from the weather database (Energyplus). The source of its data is the Egyptian Typical Meteorological Year (ETMY), developed from data provided by the U.S. National Climate Data Center.

3.3 The base case energy consumption

The base case results on the summer design date (see Table 2) were a cooling energy intensity (EUI) of 47.43 kWh/m^2 on 15 Aug, a cooling system design capacity of 62.41 kW and an annual energy cooling load of 60.758 kWh. The highest radiant temperature recorded was 33.97°C. The radiant temperature of the summer design date is shown in Fig. 7.

Table 2: Base case total cooling loads, design capacity and EUI.

	Annual energy cooling load (kWh)	Design capacity (kW)	Cooling energy use intensity (EUI) (kWh/m^2 yr)
Base case	60.758	62.41	47.43

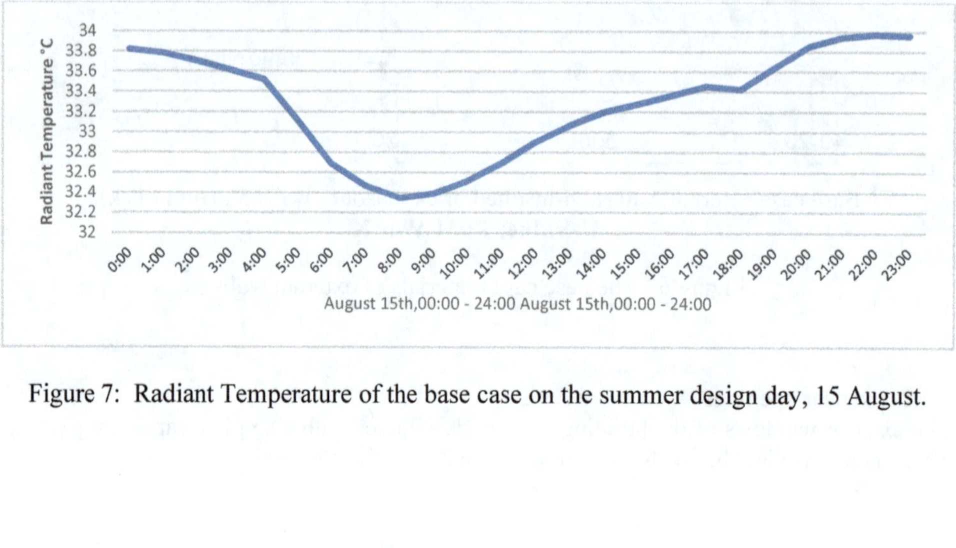

Figure 7: Radiant Temperature of the base case on the summer design day, 15 August.

4 RESULTS AND DISCUSSION

Different interventions were studied to estimate the impact of nanomaterials on the building's energy-saving; each was examined individually to determine its effect on the building's energy consumption.

The results are presented by the energy-intensity (EUI) values of the cooling energy for each simulated intervention and then calculate the percentage of energy-saving achieved compared to the base case. EUI is calculated for the building by dividing the total annual energy consumption by the entire floor space with (kWh/m^2 yr) unit measurement. The lower EUI, the better the energy performance [10].

4.1 Nano insulating materials for external walls

The vacuum insulation panel (VIP) is one type of super insulation material (SIM). The VIPs with expanded polystyrene (EPS) cover layers (see Fig. 8) were selected for thermal insulation of the building's envelope, where the EPS cover layers were applied to decrease the thermal bridging produced by the VIP metal envelope. Therefore, the applied insulation layers consist of 20 mm EPS, 20 mm VIP, and 20 mm EPS. The VIPs' thermal conductivity is 0.002–0.008 W/(m.K), and the specific heat capacity is 850 J/(kg.K) [11].

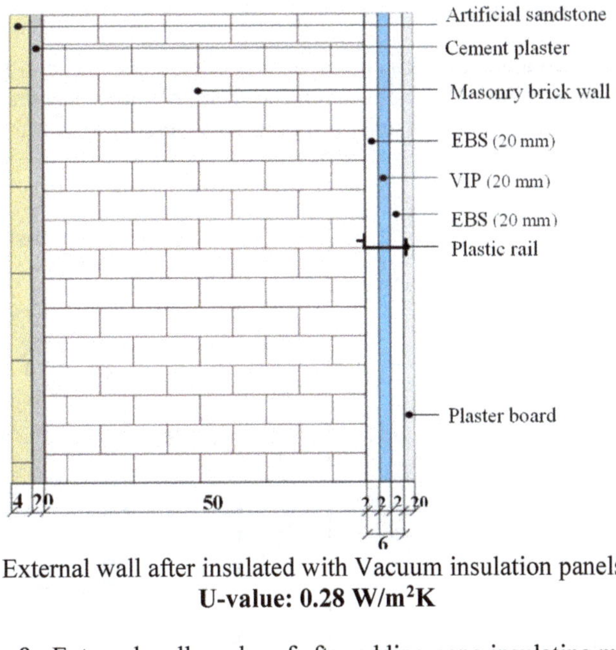

External wall after insulated with Vacuum insulation panels.
U-value: 0.28 W/m^{2}K

Figure 8: External walls and roof after adding nano insulating material.

4.1.1 Results of the VIPs external wall insulation.

The energy use of the base case windows was 47.43 kWh/m^2/year for cooling. The results (see Fig. 9) of adding VIPs to the external wall insulation showed that total energy use was reduced by more than 4.9%. The difference in the percentage reduction in energy use is insignificant because the original building walls were made of 500 mm brick wall thickness with good thermal insulation (U-value = 1.031 W/m²K). Besides, the reliance of many building areas on natural ventilation takes advantage of the mild climate of Alexandria during some periods of the summer day. Therefore, VIPs were applied to the interior walls to examine their effect on the thermal insulation walls of air-conditioned areas to achieve optimum energy efficiency.

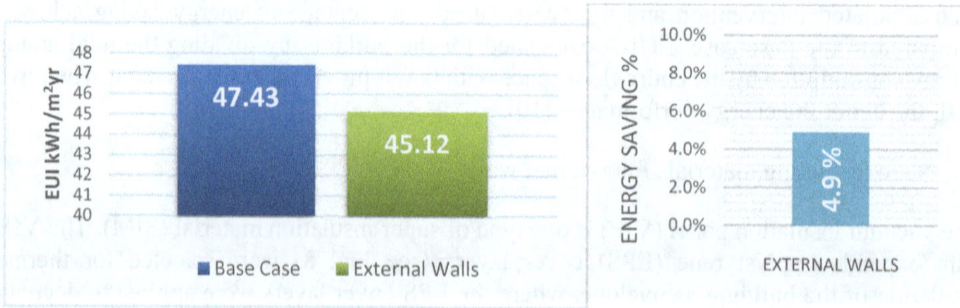

Figure 9: External walls cooling EUI, and energy saving compared to the base case.

4.2 Insulation of interior walls

Many of the building rooms are ventilated by natural ventilation, where a fan circulates the air in non-conditioned rooms to create homogenous temperature and thermal conditions. Therefore, interior insulation was assumed to be a feasible solution to achieve acceptable energy performance.

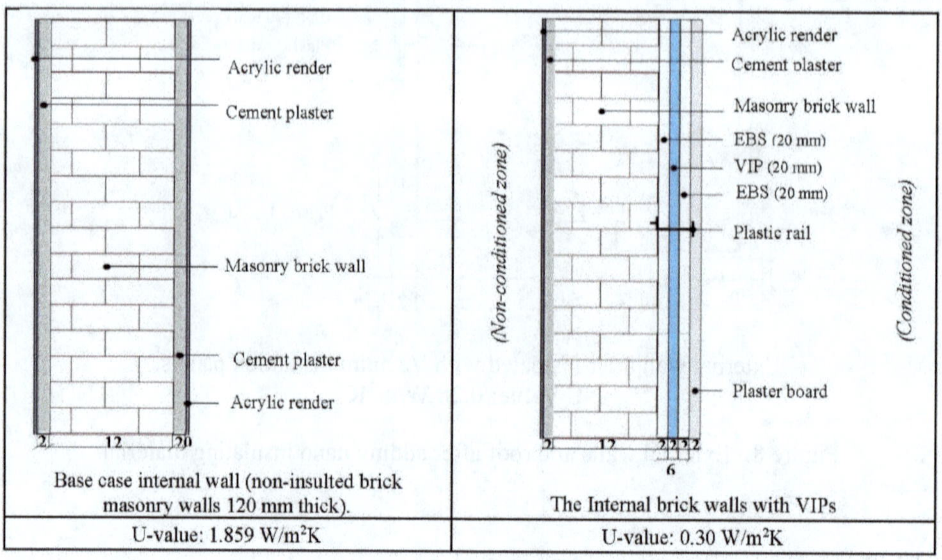

Figure 10: Internal thermal insulation for external and internal walls.

4.2.1 Results of the internal walls insulation:
The interior walls of the building are 120 mm non-insulated brick walls with a U-value of 1.859 W/m^{2}K; applying a 10 mm VIPs interior insulation granted a U-value of 0.30 W/m^{2}K. Fig. 11 shows the importance of adding thermal insulation to the interior and exterior walls; adding VIPs to internal walls alone saved about 8.9% of the energy use.

In a building with conditioned and non-conditioned areas, interior walls must be thermally insulated to keep the conditioned areas preserving their temperature during cooling and preventing heat from transferring through their walls, which decreases the use of conditioning systems and reduces energy consumption.

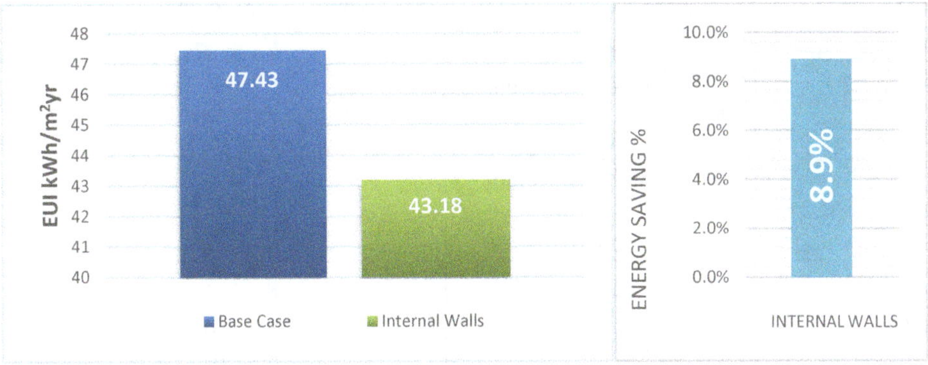

Figure 11: Internal walls cooling EUI, energy saving compared to the base case.

4.3 Thermal insulation glazing

This stage investigated the impact of nano-glazing materials on the studied building energy consumption. Table 3 shows the studied glazing characteristics.

4.3.1 Results of the glazing cases
Simulation results (see Fig. 12) revealed various improvements in the cooling EUI, with the highest improvement percentage of approximately 21% for the (G5) window, the double-glazed with a nanogel layer and two layers of argon. It is higher than the double-glazed with a 10 mm nanogel layer (G4), achieving about 9% saving. At the same time, the 6 mm single-glazed window (G1) scored a minimal impact on cooling energy intensity by 1.68%. The doubled glazing with the gab of air (G2) and void fill with argon (G3) achieved 3.12% and 4.02% savings in cooling energy intensity.

Compared to argon and air, the low thermal conductivity of aerogel allows for exceptional thermal insulation of nanogel glazing. In addition, the lower solar heat gain coefficient and U-value of aerogel increased its heat transfer [12].

4.4 All Nano insulating materials interventions in building's walls and windows

All the interventions were applied to the final simulation to determine the total energy saving. Accordingly, the double-glazed with a nanogel layer and two layers of argon was chosen as it achieved the most significant improvement. VIPs were used on the external walls and conditioned areas' internal walls.

Table 3: Specifications of the windows.

Symbol	Window glazing	Solar transmission (SHGC)	Visible light transmission	U-value (W/m^2K)
G0 (Base case)	Single clear 3 mm	0.858	0.837	6.257
G1	Single clear 6 mm	0.810	0.881	6.121
G2	6 mm clear glass, 13 mm air gap, 6 mm clear internal glass	0.697	0.781	2.708
G3	6 mm clear glass, 13 mm void filled with argon, 6 mm clear internal glass	0.698	0.781	2.549
G4	6 mm clear glass, 10 mm nanogel, 6 mm clear internal glass	0.52	0.379	2.061
G5	6 mm clear glass, 13 mm Argon, 10 mm nanogel, 13 mm Argon, 6 mm clear internal glass	0.35	0.31	0.45

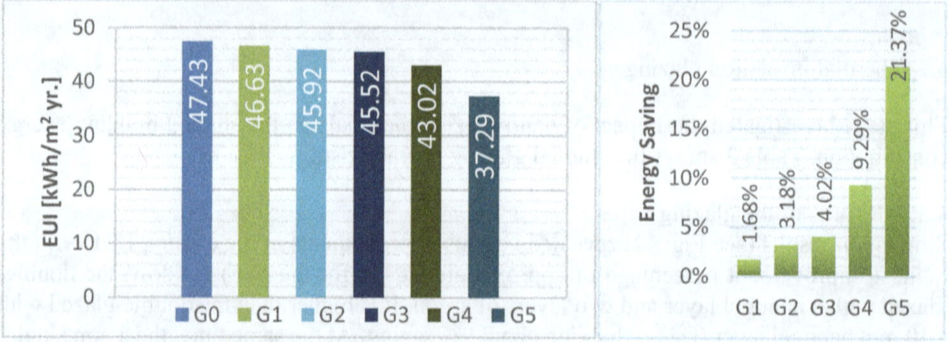

Figure 12: Windows glazing's cooling energy intensity (EUI) and energy saving percentage compared to the base case.

4.4.1 Results of nano insulating materials in walls (external and internal) and windows

The results (see Fig. 13) demonstrated an apparent reduction of 36.5% in cooling energy use intensity (EUI) for internal walls and windows thermal insulation relative to the base case.

In addition, the total cooling energy consumption decreased from 60.758 to 54.231 kWh, with a decrease in cooling loads by 6528 kWh annually, which is approximately 10.7% of the entire cooling loads. While the highest radiant temperature recorded was 31.02°C which, in turn, decreased by 2.95°C (see Fig. 14), as it recorded 33.97°C in the base case without any interference from HVAC systems.

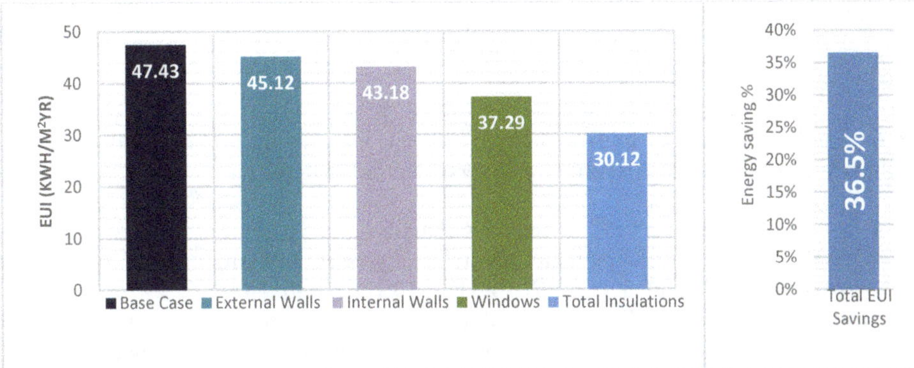

Figure 13: The cooling energy intensity (EUI), and energy-saving of all retrofitting interventions in building walls and windows.

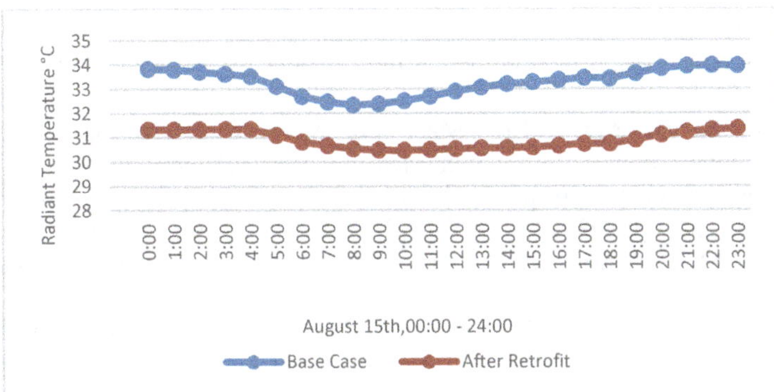

Figure 14: Comparison of the radiant temperature of the base case and after retrofitting on the summer design day, 15 August.

5 CONCLUSION

The present study investigates two nano-insulating materials – VIP and nanogel (aerogel) – to examine their potential for existing buildings, where the reduction of energy consumption and improved thermal comfort in buildings is essential. The energy use intensity (EUI), annual energy consumption, and energy savings of an educational building in Alexandria city, Egypt, were assessed.

The main outputs of the present study can be outlined as follows:

- Up to 21.37% improvement in total cooling energy use intensity (EUI) is achieved when nanogel glazing is used in the windows.
- Up to 4.9% and 8.9% improvement in total cooling, energy use intensity (EUI) is achieved when VIPs are internally used in the external and internal walls of an existing building.

These results can be used as a guideline for architects, designers and owners of buildings to examine advanced materials and to predict the energy performance for existing buildings. Furthermore, to utilize building energy simulation software as an aiding tool to indicate the building's performance and estimate different solutions to improve building energy efficiency.

ACKNOWLEDGEMENTS

The first author is grateful to Dr. Zeyad El-Sayad, and all the architectural department members and colleagues at Alexandria University for making the current study achievable.

REFERENCES

[1] Casini, M., Nano insulating materials and energy retrofit of buildings. *AIP Conference Proceedings*, **1749**(1), 020005, 2016. DOI: 10.1063/1.4954488.

[2] Johansson, P., Donarelli, A. & Strandberg, P., Performance of insulation materials for historic buildings: Case studies comparing a super insulation material and hemp-lime. *3rd International Conference on Energy Efficiency in Historic Buildings (EEHB2018)*, Visby, Sweden, Uppsala University, pp. 80–88, 2018.

[3] Gonçalves, M., Simões, N., Serra, C. & Flores-Colen, I., A review of the challenges posed by the use of vacuum panels in external insulation finishing systems. *Applied Energy*, **257**, 114028, 2020. DOI: 10.1016/j.apenergy.2019.114028.

[4] Casini, M., Smart buildings: Advanced materials and nanotechnology to improve energy-efficiency and environmental performance. Woodhead Publishing, pp. 149–159, 2016.

[5] Ascione, F., De Masi, R.F., Mastrullo, R.M., Ruggiero, S. & Vanoli, G.P., Experimental investigation and numerical evaluation of adoption of multi-layered wall with vacuum insulation panel for typical *Mediterranean Climate, Energy and Buildings*, **152**, pp. 108–123, 2017. DOI: 10.1016/j.enbuild.2017.07.029.

[6] Mujeebu, M.A., Ashraf, N. & Alsuwayigh, A., Energy performance and economic viability of nano aerogel glazing and nano vacuum insulation panel in multi-story office building. *Energy*, **113**, pp. 949–956, 2016. DOI: 10.1016/j.energy.2016.07.136.

[7] Buratti, C. & Moretti, E., Glazing systems with silica aerogel for energy savings in buildings. *Applied Energy*, **98**, pp. 396–403, 2012. DOI: 10.1016/j.apenergy.2012.03.062.

[8] Fouad, F., Towards a green university: The green campus of Engineering Faculty Alexandria University. PhD thesis, Alexandria University, Egypt, p. 72, 2017.

[9] Saleem, A.A., Abel-Rahman, A.K., Ali, A.H.H. & Ookawara, S., An analysis of thermal comfort and energy consumption within public primary schools in Egypt. *IAFOR Journal of Sustainability, Energy and the Environment*, **3**, 2016.

[10] Garg, V., Mathur, J. & Bhatia, A., *Building Energy Simulation: A Workbook Using Designbuilder™*, CRC Press, 2020. DOI: 10.1201/9780429354632.

[11] Quenard, D., Long-term performance of super insulating materials in building components and systems-IEA-EBC Annex 65. *Proceedings of International Conference CISBAT 2015 Future Buildings and Districts Sustainability from Nano to Urban Scale, LESO-PB, EPFL*, pp. 125–130, 2015.

[12] Moretti, E., Zinzi, M., Merli, F. & Buratti, C., Optical, thermal, and energy performance of advanced polycarbonate systems with granular aerogel. *Energy and Buildings*, **166**, pp. 407–417, 2018. DOI: 10.1016/j.enbuild.2018.01.057.

SECTION 3
OCCUPANT AND USER BEHAVIOUR

KEY FACTORS OF OCCUPANTS' BEHAVIOUR CHARACTERISTICS AND INDOOR AIR QUALITY PERCEPTION ON OCCUPANCY COMFORT IN MULTI-STOREY RESIDENTIAL BUILDINGS IN THAILAND

WARANGKANA JUANGJANDEE, VICKI STEVENSON & ESHRAR LATIF
Welsh School of Architecture, Cardiff University, UK

ABSTRACT

Multi-storey residential buildings in high-density urban areas, especially in a tropical climate like Thailand, tend to suffer poorer indoor environmental quality. Such a significant impact increases the risks of health issues. However, there has been little study exploring the reason and identifying potential solutions. This research tackles the problems of thermal comfort and indoor air quality (IAQ) in multi-storey residential buildings in high-density urban areas (2,400 people/km^2) in Chiang Mai, Thailand (equatorial, winter dry climate). The survey explored occupants' behaviour characteristics in relation to their indoor air quality and comfort perceptions. 482 responses from adult occupants were received (400 responses were required for results to have a 95% confidence level). Cronbach's alpha, Spearman's rank, Pearson correlation coefficient and stepwise multiple regression analysis methods were applied to analyse data; the five key results are summarised as follows: (1) five critical factors for overall comfort satisfaction were identified as temperature, natural lighting, ventilation, room privacy, and humidity (in order of importance); (2) activities which generate moisture (e.g., washing clothes, hanging clothes to dry, cooking) have a strong negative correlation with IAQ perception; (3) factors such as natural lighting and ventilation had a significant positive relationship with IAQ, while humidity and mould had a negative relationship with the IAQ; (4) natural lighting and air freshness were the strongest influences on occupant's humidity comfort; and (5) ventilation and air freshness positively correlate with ventilation comfort, whereas humidity negatively correlated with the comfort (at a significant level). This study proposes design guidelines to improve the comfort. Built environment professionals from Thailand will be consulted on the proposed design, which may assist architects and engineers in generating IAQ improvement for their occupants, achieving indoor comfort.

Keywords: occupants' behaviour, occupancy comfort, multi-storey residential building, surveying, indoor air quality, questionnaire.

1 INTRODUCTION

The rapid increase in multi-storey residential buildings in metropolitan area has caused high-density environment, which has caused an obstruction of wind flow and solar radiation over decades [1], [2]. The obstruction exacerbates both outdoor and indoor thermal comfort, especially in tropical climates [3], [4]. Many studies conducted within the last decade have shown that the limitation of opening due to the obstruction can reduce the change in ventilation flow rate and indoor air quality [5]–[7]. The rising mean internal temperature cause problems with overheating in building [8], which potentially further impacts human health risk [9] and building electricity in cooling load demanded [8]. Additionally, the ineffective building design negatively impacts living comfort by causing difficulties in removing internal moisture. The high intensity of moisture might cause mould growth to risk human health [10]. The moisture-related issues, particularly in tropical climate zones, aggravate residential buildings more than other building types [11].

The problem related to indoor environment quality is defined by the specific occupancy behaviour causing the problems and discomfort. Yan et al. [12] also stated that occupant

www.witpress.com, ISSN 1743-3509 (on-line)
doi:10.2495/ARC220061

behaviour mainly influences building performance. The behaviour pattern can evolve concerning a mix of sources, i.e., inside-gender, preference, and outside-air temperature, wind speed, building properties [13]. The mix of sources can impact indoor environment conditions. For example, Roetzel et al. [14] investigated that windows are usually opened to improve natural ventilation and enhance fresh air entering the room. However, humid air can also enter the room through the window and crack in a free-running building.

Thus, understanding the occupant's room and behaviour characteristics is crucial for the building evaluation. A survey is one of the methods for obtaining feedback on a building's performance in use after it has been occupied. Xue et al. [15] adopt the Chi-Square and non-parametric test, i.e. Spearman rank coefficient to identify the relationship between occupants' behaviour and luminous comfort in residential buildings; the result indicated that occupants' behaviour significantly influenced luminous comfort. Azimi and Esmaeilzadeh [16] applied Cronbach alpha for internal consistency testing. Yik et al. [11] adopted survey and physical measurement for investigating moisture generation due to various household activities. The researcher mentioned that survey could be applicable to define actual activities and moisture generation in the household. Unfortunately, there have been limited studies on occupants' behaviour with their building context, perception, and comfort.

This study investigates the critical factors of occupants' behaviour characteristics and indoor air quality perception on occupancy comfort in a multi-storey residential building in Chiang Mai, Thailand. Furthermore, this study visualises the satisfactory and environmental performance of the residential buildings with occupants living in. The online questionnaire constituted multiple-choice and Likert scale questions. The occupants are encouraged to select the most relevant answer about general background, occupants' behaviour, and physical environment, through the multiple-choice questions. Then, the participants rate their room's environmental perception through Likert scale questions.

2 METHODOLOGY

2.1 Sampling and questionnaire design

Chiang Mai is selected as a case study. Chiang Mai is a tropical savanna climate (Aw) located in the north of Thailand. Chiang Mai is the fifth largest population city across Thailand, with a population density above 1,200 people per km^2 in the urban area [17].

This survey was carried out from February to May 2021. One criterion applies to the participants; only adults (age range 18–60) live in the multi-storey residential buildings in Chiang Mai. The minimum number of necessary samples to meet the desired statistical constraints (confidence level with 95%) is 400 from the Chiang Mai metropolitan population of 190,199 people [18], only those living in multi-storey residential buildings. Yamane's formula for calculating sample size is:

$$\text{Sample size (n)} = N/(1 + N(e\hat{}2)), \tag{1}$$

where n, N, and e are the sample size, population size and level of precision, respectively [19]. Simple random sampling was conducted to select two districts (clusters) in Chiang Mai metropolitan. Then, the investigation used the cluster sampling method with Thai census data sources, who live in multi-storey residential buildings in Chiang Mai, for choosing samples. The questionnaire invited approximately 4,000 people to participate because the average online questionnaire response rate is approximately 10%–15% of sample [20]. 482 completed surveys were obtained, meeting the requirement to have a 95% confidence level in the results.

The questionnaire has been approved by the ethics group in Cardiff University.

The questionnaire was divided into five parts. The first part is general information of participant. Part 2 considered the physical living environment. Part 3 considered occupancy indoor activity, including cooking, and washing clothes characteristics. The questions in parts 1 to 3 were multiple choice. Part 4 included questions about the participants' perceptions towards IAQ. The questions required the residents to answer on a nine-point scale. Part 5 considered the satisfactory level of the participants with the environment in their rooms.

2.2 Data analysis methods

All data were analysed via SPSS 27.0. For reliability measuring, Cronbach's alpha was used to assess the internal consistency of tested items [21], such as occupants' comfort questions and satisfaction with indoor air quality levels.

Descriptive analysis was used to evaluate general information of occupants, while inferential statistics was used to analyse the relationship between occupants' room perception and satisfaction. In order to investigate the relationship between two variables, Spearman's rank correlation was applied to measure the monotonic association (between ordinal scales) [22], while Pearson correlation was performed to measure the linear relationship (between ratio/interval scales) [23]. Both Pearson and Spearman correlation coefficients are measured on a scale varying between –1 and 1. The correlation coefficient greater than zero signifies a positive relationship, while the value less than zero indicates a negative relationship. Zero value means no correlation [24]. For the Likert scale, a correlation value close to –1 or 1 indicates a strong relationship. The correlation data having the coefficient value either lower than –0.5 or higher than 0.5 were then tested by multiple regression. Multivariance regression analysis was to identify the most significant variables influencing the outcome.

3 RESULTS AND ANALYSIS

This study used Cronbach's alpha coefficient for estimating internal consistency of occupants, indoor air quality scale and comfort in the survey. Cronbach's alpha of nine comfort questions were all above 0.60, which can be considered as good internal consistency and strong correlation in questionnaire items.

3.1 Demographic analysis of survey participants

50.8% of the respondents were female, 44.6% were male, 3.9% preferred not to answer, and 0.6% were other. Over half of respondents (60.8%) were students. For ownership, most of the participants were tenants (95%), while 5% were room owners. In terms of occupied periods, the participants commonly spent time in their rooms between 8–12 hours and 12–16 hours during the weekend (see Fig. 1).

3.2 Physical environment

Most of the participants (30%) lived in seven floors building, while approximately 19% of the participants were evenly distributed between four, five and six floor buildings. 78% of the participants occupied on the 1st, 2nd and 3rd floor. The occupants of shorter buildings

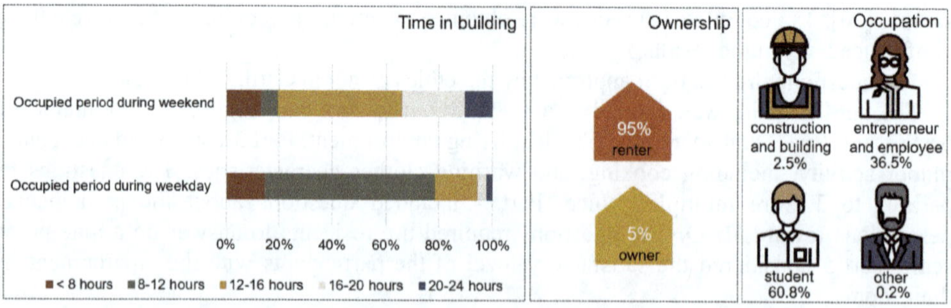

Figure 1: Demographic characteristics of participants (© Warangkana Juangjandee, 2022).

(one, two or three floors) tended to live on the 1st floor. Greater than 60% of participants lived in buildings located 0–6 m away from the nearest nearby building. In addition, almost half of the participants' room openings faced north. For building plan and room location, over 90% of the respondents lived in double load corridor buildings and over 70% of their rooms' locations were at the middle of the building. Most responders reported rooms comprising of one bedroom, one bathroom and one balcony. Three typical room types were defined in relation to location of WC. The proportion of people living in these room types (along with room sizes) is shown in Fig. 2. Over half of the respondents lived in room sizes ranging from 21 to 30 m^2. For room size range from 8–20 m^2 to 21–30 m^2, there is an even distribution of participants' room types. For example, the 21–30 m^2 size range has approximately 16% of Type A, 23% of Type B and 23% of Type C.

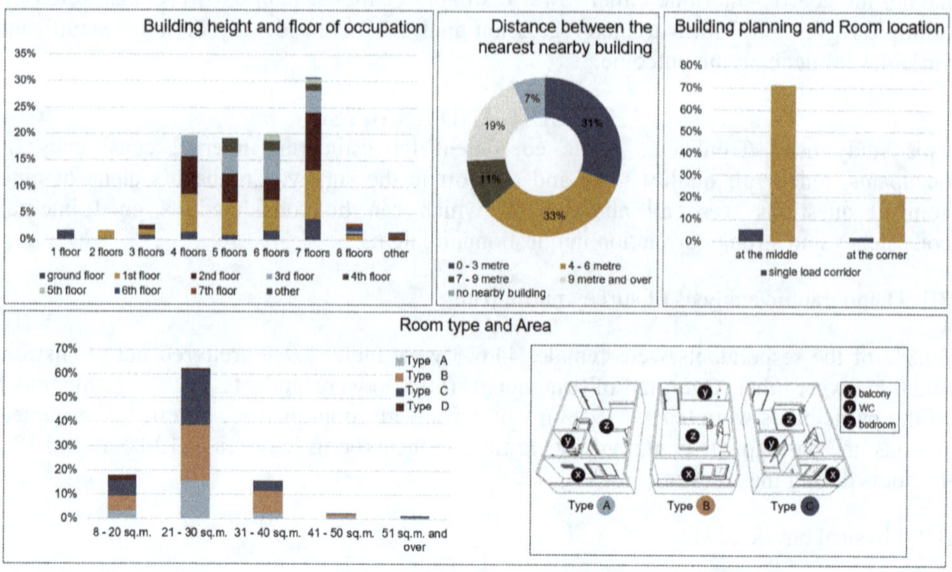

Figure 2: Physical environment (© Warangkana Juangjandee, 2022).

94.1% and 69.2% of the participants had wood for room doors and furniture, respectively. Over half of the respondents' rooms had concrete as a material of wall and ceiling. 80.9% of the voters occupied rooms that had tile as a flooring material. Plastic was a toilet door material of around 56% of the respondents. Referring to the different material porous structure, each material has different thermal properties, which will affect ventilation, moisture transport and heat transfer in building (see Fig. 3).

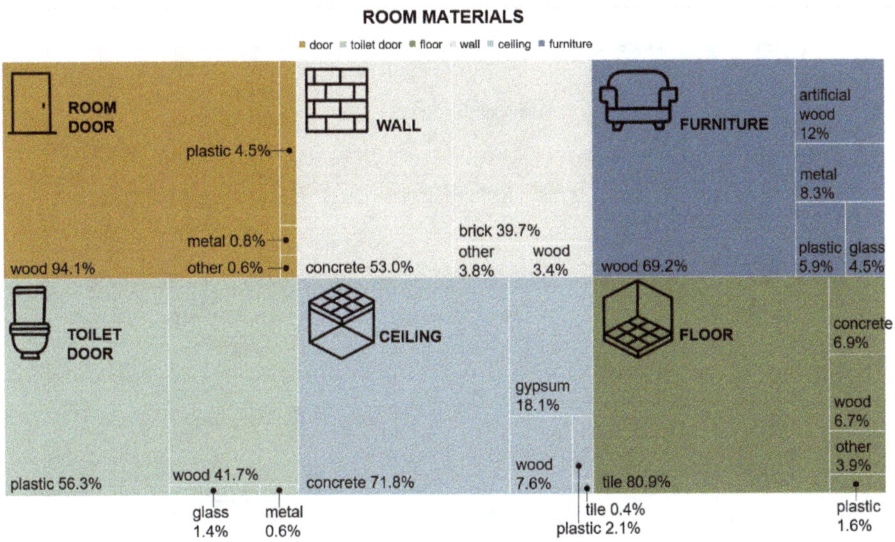

Figure 3: Treemap of occupants' room materials (© Warangkana Juangjandee, 2022).

3.4 Indoor air quality perception level scale and occupants' satisfaction

Fig. 4 presents occupants' indoor air quality perception levels, in which the temperature (a1), natural lighting (a2) and air humidity (a3) perception levels are exhibited in diverging stack bar charts; the air freshness (b1), ventilation (b2) and mould (b3) perception levels are shown in stacked bar charts.

The respondents were asked to rate their seasonal temperature perceptions. In summer, the percentage of participants indicating temperature perception of "excessively hot" is four times higher during summer than winter. This is likely to be attributable to seasonal temperature variations as the summer is approximately 5°C hotter than winter in Chiang Mai. The average outdoor temperature in summer ranges from 26.4°C to 32.0°C.

The natural lighting perception was divided into two purposes: the lighting for general and working purpose. Approximately 50% of the respondents perceived natural lighting rather dark for general purpose. Moreover, over 60% of the participants felt that the natural lighting for working purposes was not bright enough. Over half of participants responding to air humidity and ventilation report "excessively humid" and "not significant", respectively. Over half of the participants reported stale air in their rooms. The surface with the relative humidity exceeding 80% can support gemination of various mould spores. In addition, in warm, humid air within the building, the excess moisture on the surface can be accumulated as condensation, which can be a risk of construction damage. The percentage of mould growth in the room can represent the actual room environment. 55.2% of the

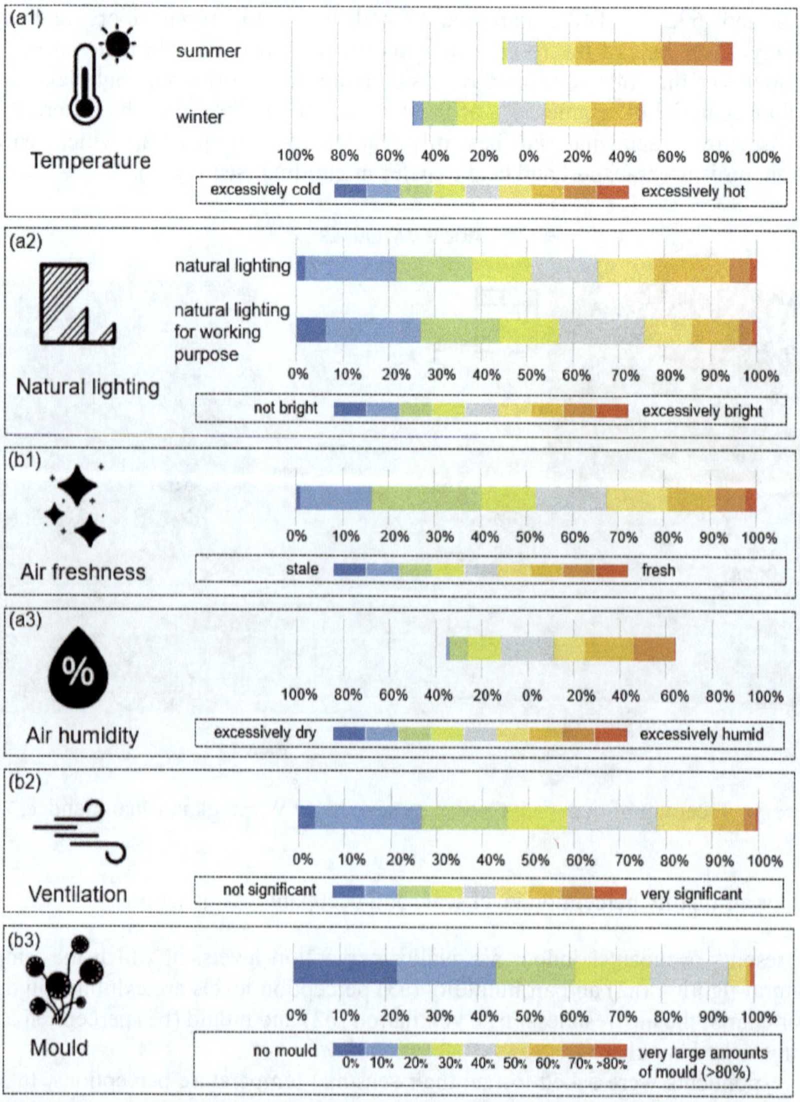

Figure 4: Indoor quality perception level (© Warangkana Juangjandee, 2022).

respondents had mould happening on between 10% and 30% of their rooms' surface areas, whereas a further 20% of the respondents had mould occurring on more than 40% of the rooms' surface.

For occupant satisfaction, the proportion of participants who agreed with the statements are shown to the right of the zero line in the bar chart, while the participants who disagreed are shown to the left (in Fig. 5). Across the categories, approximately 30% of respondents gave a neutral response, while the average satisfaction response was roughly 20%. However, a considerable number of votes indicated dissatisfaction. The percentage of dissatisfaction in natural lighting, temperature, location and room size was over 40%, while the dissatisfaction was 38% and 28% in room humidity and natural ventilation, respectively.

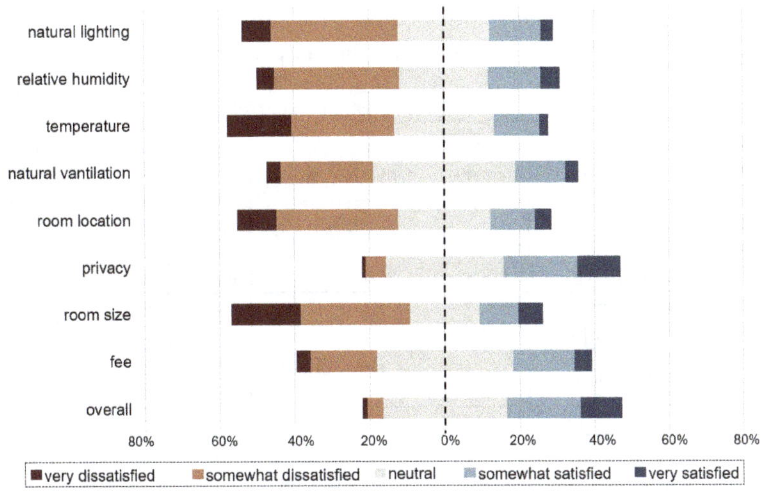

Figure 5: Occupants'satisfaction (© Warangkana Juangjandee, 2022).

3.5 Relationship between indoor air quality scale and occupants' satisfaction

This section used two steps to examine the relationship between indoor air quality perception and occupant satisfaction. Firstly, Pearson's correlation coefficient was applied to define the strength of any linear relationship between two variables. Secondly, multiple regression was used to determine the main contributing factors in the relationship between variables.

The occupants' satisfaction involved two primary aspects, which are the physical environment and indoor air quality perception. Table 1 presents all the factors which have a correlation significant at the 0.01 level to indoor air quality scale. The relationship between human behaviour and indoor quality rating was estimated via Spearman correlation, while the Pearson's correlation was applied to assess the relationship between ratio scale, e.g., building height and natural lighting perception. Referring to Cohen rules, a correlation coefficient (reported as the statistic r) of 0.1 refers to a weak correlation, 0.3 means a moderate correlation and 0.5 means a strong correlation. The p-value (quoted under significant (2-tailed) at the 0.01 level ($p < 0.01$)) is significant if the value is between 0.001 to 0.010, which also indicates 99% confidence levels. The p-value between 0.010 to 0.050 will be considered significant at 95% confidence level [26].

Human behaviour characteristics such as cooking, washing clothes, and hanging clothes for drying had a strong significant correlation ($p < 0.01$) with indoor air quality perception. For example, the washing characteristic had a solidly negative correlation with the mould rating scale, which concludes mould is likely to increase if washing activities happen in the room. The correlations between the physical environment and indoor air quality scale were also strong, with significant at the 0.01 level. The coefficients between the distance from the participant's building to the nearby building were 0.509 with natural lighting, 0.530 with fresh air scale, and 0.542 with ventilation scale. This can signify that such a positive relationship had large strength of association because the coefficients were above 0.5. However, the distance between the participant's building and a nearby building was

Table 1: Spearman and Pearson correlation coefficient of human behaviour characteristic, physical environment and indoor air quality perception scale.

Spearman correlation			Pearson's correlation		
Cooking characteristic	Air freshness	–.510**	Cleaning frequency	Humidity satisfaction	–.212**
	Air humidity	.563**		Temperature satisfaction	–.245**
	Ventilation	–.504**		Room privacy satisfaction	–.157**
	Mould level	.499**		Overall satisfaction	–.318**
Washing characteristic	Air freshness	.582**	Room size	Room size satisfaction	.566**
	Air humidity	–.670**	Building tall	Natural lighting	–.363**
	Ventilation	.548**		Natural lighting for working purpose	–.368**
	Mould level	–.617**	Distance between the building and nearby building	Natural lighting	.509**
Cloth hanging characteristic	Air freshness	.364**		Natural lighting for working purpose	.478**
	Air humidity	–.433**		Air freshness	.530**
	Ventilation	.374**		Air humidity	–.601**
	Mould level	–.411**		Ventilation	.542**
				Mould level	–.417**

** Correlation is significant at the 0.01 level (2-tailed).
* Correlation is significant at the 0.05 level (2-tailed).

negatively correlated with the air humidity air scale. The result indicates that greater distance between the participant's building and nearby building can influence the air humidity perception in a room.

The overall satisfaction had significant relationship ($p < 0.01$) with indoor quality scale including temperature, ventilation, humidity, and mould level. In Table 2, humidity satisfaction increased if the room has higher natural lighting, greater fresh air, less wet air, more ventilation, and less mould level. Additionally, higher fresh air, less wet air, more ventilation and slight mould level enhance ventilation satisfaction level. Most of the satisfaction categories (except location) had significant association with overall satisfaction (p-value < 0.1).

After the correlation analysis, the strong significant relationship is analysed stepwise by multiple regression for further finding the main factors on occupants' satisfaction. Regression is a statistical technique to model the relationship between a single dependent variable and multi-independent variables. For estimating the coefficients in the regression equation, unstandardised value (B) shows the influence of independent variables on occupants' satisfaction. The larger value of B is considered as the larger influencer on such a dependent factor. p-value is the significance level of the regression, and t-value is used to measure a coefficient's standard error, in which t-value closing to zero indicates the low reliability of the predictive power on its coefficient [27]. Occupant's satisfaction was determined as the dependent variable, and other significant factors were set as independent variables.

Most independent variables of humidity satisfaction are significant at 0.05 level (p-value < 0.05), except natural lighting for working purpose. Natural lighting ($B = 0.146$, $t = 3.948$) and air freshness level ($B = 0.143$, $t = 4.849$) are principal influences on humidity satisfaction at 99% confidence levels, respectively. Natural lighting relates to humidity

Table 2: Pearson correlation of indoor quality perception scale and occupants' satisfaction.

Pearson's correlation			Pearson's correlation		
Natural lighting satisfaction	Temperature in winter	–.438**	Ventilation satisfaction	Air freshness	.737**
	Temperature in summer	–.218**		Air humidity	–.701**
	Natural lighting	.783**		Ventilation	.781**
	Natural lighting for working purpose	.757**		Mould level	–.605**
Humidity satisfaction	Natural lighting	.688**	Room privacy satisfaction	Natural lighting	.536**
	Natural lighting for working purpose	.650**		Natural lighting for working purpose	.533**
	Air freshness	.688**		Air freshness	.544**
	Air humidity	–.621**		Air humidity	–.500**
	Ventilation	.666**		Ventilation	.575**
	Mould level	–.645**	Overall satisfaction	Temperature in winter	–.529**
Temperature satisfaction	Temperature in winter	–.515**		Temperature in summer	–.352**
	Temperature in summer	–.550**		Natural lighting	.579**
	Air freshness	.493**		Natural lighting for working purpose	.582**
	Air humidity	–.339**		Air freshness	.622**
	Ventilation	.540**		Air humidity	–.490**
	Mould level	–.342**		Ventilation	.644**
				Mould level	–.482**

** Correlation is significant at the 0.01 level (2-tailed).
* Correlation is significant at the 0.05 level (2-tailed).

satisfaction because natural lighting can influence human subjective thermal perception from the psychological perspective [28]. Moreover, more natural lighting can associate with more solar radiation, which also effects human thermal perception. The relationship between natural lighting for working and humidity satisfaction is statistically significant at 0.5 level (p-value > 0.5). Natural lighting for working purpose is not able to predict or influence on humidity satisfaction, while other factors have influenced on the humidity satisfaction with significant at 0.05 level (p-value < 0.05).

Three factors influencing the ventilation satisfaction. Ventilation level is a major factor (B = 0.265, t = 9.861, p-value < 0.001), following by air humidity perception level (B = –0.205, t = –8.708, p-value < 0.001) and air freshness (B = 0.133, t = 5.469,p-value < 0.001), respectively.

The main influence of room privacy satisfaction is ventilation rate (B = 0.190, t = 4.118, p-value < 0.001). Air humidity and air freshness perception level are minor factors on the satisfaction. Ventilation perception correlates to privacy satisfaction because rooms with more privacy can open their rooms' opening and then benefit to more ventilation in rooms. However, natural lighting in general and for working purpose are not statistically significant because p-values are greater than significant level (0.05).

The multi-linear regression was calculated to predict the overall satisfaction based on indoor their environmental satisfaction. The coefficient of regression is shown in Table 3. Temperature satisfaction is a major influence on overall satisfaction (B = 0.199, t = 6.163, p-value < 0.001). Natural lighting (B = 0.172, t = 4.803, p-value < 0.001) and ventilation satisfaction (B = 0.159, t = 4.426, p-value < 0.001) are the second and third influences on the overall satisfaction, respectively.

Table 3: Coefficients of regression for overall satisfaction.

Dependent variable	Independent variable	Unstandardized coefficients (B)	t	p-value
Overall satisfaction	(Constant)	0.896	9.907	0.000
	Natural lighting satisfaction	0.172	4.803	0.000
	Humidity satisfaction	0.080	2.358	0.019
	Temperature satisfaction	0.199	6.163	0.000
	Ventilation satisfaction	0.159	4.426	0.000
	Room privacy satisfaction	0.153	6.251	0.000

4 DISCUSSION

Physical environment, such as distance between occupants' building and nearby buildings and building height, affects indoor air quality perception. For example, occupants are likely to perceive inadequate natural lighting and low natural ventilation when another building is located between 0–3 m from their buildings. In contrast, the occupants perceive air rather humid when there is a smaller distance between the occupant's building and nearby building. Referring to Yan et al. [12] and Foster and Oreszczyn [29], occupants in a building with a short distance between nearby buildings tend to keep privacy by keeping the blind down, which can block natural ventilation entering the room. Based on the result of the study, it is suggested that buildings having short distance to nearby buildings should be designed whereby the building or building system for maintaining privacy and also sustaining an adequate indoor air quality of living.

Occupant activities generating moisture content in the room, such as cooking, washing and hanging cloth for drying, influence the occupants to perceive stale air, high humid air, low ventilation and high level of mould occurring. An earlier survey by Yik et al. [11] and Zemitis et al. [30] stated that moisture is generated in residence from occupant not only via internal occupant's body but also by various household activities such as cooking, clothes hanging, clothes washing etc. Occupancy activities, room function, usage duration and frequency cause different level of moisture generation in room. For example, 2,000–3,000 g, 1,500 g, 500 g per day for cooking, drying clothes indoor and washing clothes respectively [25]. In tropical region, the building should be dedicatedly designed by minimizing indoor moisture accumulation. However, from the survey results (Fig. 4), most building materials are porous materials, such as wood door, brick wall and wood furniture, that are susceptible to moisture damage.

Relationship between indoor air quality scale and occupants' satisfaction found that temperature satisfaction is the primary influence on overall satisfaction. The result confirms Tanabe et al. [9] mentioned that temperature affects human mental and body perception. In addition, a high ventilation flow, adequate natural lighting, and low air humidity are also crucial to achieving room satisfaction (see Table 2). Poor natural ventilation is reported as a significant challenge to residential occupancy, which increases the chance of indoor mould growth and increases sick building syndromes [31]. The limitation of natural lighting in buildings can affect occupants' visual and atmosphere perception [32]. !

5 SUMMARY

The survey was analysed by statistical method. The conclusions about the factors influencing occupants' indoor environmental quality perception and their satisfaction are listed below:

1. The most popular physical environment of participants' rooms will be used as a base case in CFD simulation in further steps, such as type B of room planning, room materials, and building height.
2. Occupant indoor air perception has a significant influence on room satisfaction. The occupants' ventilation satisfaction is mainly influenced by ventilation, air humidity and air freshness perception, respectively. The humidity satisfaction is also determined by those perceptions as well as indoor air quality perception of natural lighting and mould level.
3. The environmental satisfaction of natural lighting, humidity, temperature ventilation and room privacy are the five key factors of the overall comfort satisfaction. In such vital factors, temperature satisfaction is the primary factor of the overall satisfaction, followed by natural lighting and ventilation, respectively.

The study examines the data influencing occupants' satisfaction to be a guideline for developing a design to improve occupants' ventilation room and achieve room satisfaction, especially in the tropical climate.

ACKNOWLEDGEMENT

This work is supported by a grant from Energy Policy and Planning Office, Ministry of Energy, Thailand.

REFERENCES

[1] Arnfield, A.J., Two decades of urban climate research: a review of turbulence, exchanges of energy and water, and the urban heat island. *International Journal of Climatology,* **23**(1), pp. 1–26, 2003.

[2] Garcia-Nevado, E., Pages-Ramon, A. & Coch, H., Solar access assessment in dense urban environments: The effect of intersections in an urban canyon. *Energies,* **9**(10), p. 796, 2016.

[3] Rodriguez, C.M. & D'Alessandro, M., Indoor thermal comfort review: The tropics as the next frontier. *Urban Climate,* **29**, 100488, 2019.

[4] Lau, K.K.-L., Chung, S.C. & Ren, C., Outdoor thermal comfort in different urban settings of sub-tropical high-density cities: An approach of adopting local climate zone (LCZ) classification. *Building and Environment,* **154**, pp. 227–238, 2019.

[5] Khan, N., Su, Y. & Riffat, S.B., A review on wind driven ventilation techniques. *Energy and Buildings,* **40**(8), pp. 1586–1604, 2008.

[6] Karava, P., Stathopoulos, T. & Athienitis, A.K., Wind-induced natural ventilation analysis. *Solar Energy,* 81(1), pp. 20–30, 2007.

[7] Aflaki, A., Mahyuddin, N., Mahmoud, Z.A.-C. & Baharum, M.R., A review on natural ventilation applications through building façade components and ventilation openings in tropical climates. *Energy Buildings,* **101**, pp. 153–162, 2015.

[8] Li, C. et al., Interaction between urban microclimate and electric air-conditioning energy consumption during high temperature season. *Applied Energy,* **117**, pp. 149–156, 2014.
[9] Chartered Institute of Building Services Engineers (CIBSE), A guide, environmental design, 2006.
[10] Nitmetawong, T., Boonvisut, S., Kallawicha, K. & Chao, H.J., Effect of indoor environmental quality on building-related symptoms among the residents of apartment-type buildings in Bangkok area. *Human Ecological Risk Assessment: An International Journal,* pp. 1–15, 2019.
[11] Yik, F., Sat, P. & Niu, J., Moisture generation through Chinese household activities. *Indoor and Built Environment,* **13**(2), pp. 115–131, 2004.
[12] Yan, D. et al., Occupant behavior modeling for building performance simulation: Current state and future challenges. *Energy and Buildings,* **107**, pp. 264–278, 2015.
[13] Rinaldi, A., Schweiker, M. & Iannone, F., On uses of energy in buildings: Extracting influencing factors of occupant behaviour by means of a questionnaire survey. *Energy and Buildings,* **168**, pp. 298–308, 2018.
[14] Roetzel, A., Tsangrassoulis, A., Dietrich, U. & Busching, S., A review of occupant control on natural ventilation. *Renewable and Sustainable Energy Reviews,* **14**(3), pp. 1001–1013, 2010.
[15] Xue, P., Mak, C.M. & Cheung, H., The effects of daylighting and human behavior on luminous comfort in residential buildings: A questionnaire survey. *Building and Environment,* **81**, pp. 51–59, 2014.
[16] Azimi, N. & Esmaeilzadeh, Y., Assessing the relationship between house types and residential satisfaction in Tabriz, Iran. *International Journal of Urban Sciences,* **21**(2), pp. 185–203, 2017.
[17] Nation Statistical Office, Number of population from registration by sex, area, density, house, region and province: 2009–2018, 2020. http://statbbi.nso.go.th/staticreport/page/sector/th/01.aspx.
[18] Chiang Mai Provincial Statistical Office, Population from registration record, percentage change and density by district: 2003–2020, 2020.
[19] Israel, G.D., Determining sample size, 1992.
[20] Cleave, P., What is a good survey response rate? 2021.
[21] Barnette, J.J., Effects of stem and Likert response option reversals on survey internal consistency: If you feel the need, there is a better alternative to using those negatively worded stems. *Educational and Psychological Measurement,* **60**(3), pp. 361–370, 2000.
[22] Sedgwick, P., Spearman's rank correlation coefficient. *BMJ,* **349**, 2014.
[23] Nettleton, D., Selection of variables and factor derivation. *Commercial Data Mining,* pp. 79–104, 2014.
[24] Berman, J.J., *Data Simplification: Taming Information with Open Source Tools*, Morgan Kaufmann, 2016.
[25] British Standards Institution, *Code of Practice for Control of Condensation in Buildings*, 2011.
[26] Chetty, P., Correlation of variables in SPSS, 2015. https://www.projectguru.in/correlation-variables-spss/.
[27] Sternstein, M., Statistics: Barron's Educational Series, Barron's: Hauppauge, NY, 1996.

[28] Chinazzo, G., Wienold, J. & Andersen, M., Daylight affects human thermal perception. *Scientific Reports,* **9**(1), pp. 1–15, 2019.
[29] Foster, M. & Oreszczyn, T., Occupant control of passive systems: the use of Venetian blinds. *Building and Environment,* **36**(2), pp. 149–155, 2001.
[30] Zemitis, J., Borodinecs, A. & Frolova, M., Measurements of moisture production caused by various sources. *Energy and Buildings,* **127**, pp. 884–891, 2016.
[31] Milton, D.K., Glencross, P.M. & Walters, M.D., Risk of sick leave associated without door ventilation level, humidification, and building related complaints. *Indoor Air 10.4,* **10**(4), pp. 212–221, 2000.
[32] Lee, J.-H., Yoon, Y., Baik, Y.-K. & Kim, S., Analyses on human responses to illuminance variations for resident-friendly lighting environment in a small office. *Indoor and Built Environment,* **22**(3), pp. 535–550, 2013.

URBAN FABRICS OF HOUSING FORMS AND ITS IMPACT ON HUMAN BEHAVIOR

MONICA H. FAHIM[1,2], AHMED M. SOLIMAN[1] & DINA SAADALLAH[1]
[1]Faculty of Engineering, Alexandria University, Egypt
[2]Faculty of Engineering, Pharos University, Egypt

ABSTRACT

An individual's psychological life and their incorporation with the built environment have a remarkable relationship with urban space syntax. The scarcity of positive communication between humans and place-making is considered the core problem. This phenomenon occurs when urban designers ignore the role of humanity during the design process without taking into consideration its influence on the development of urban spaces. An individual's well-being and their quality of life are affected by the degree of residential satisfaction. Factors contributing to residential satisfaction include: (1) physical factors that could be determined in the availability of parks and facilities in the community; (2) social factors that could be determined in the feeling of belonging to a community; and (3) personal factors that could be determined in people's duration of residence and homeownership. The paper aims to examine the relations between human behavior and urban housing form through investigating the human behavior associated with the physical environment and determine the factors affecting it. The paper's assumption is that improving the quality of the space syntax will increase the residents' belonging to their built environment, in addition to enhancing their residential satisfaction. Increasing levels of sense of community, sense of belonging, and sense of place could achieve residential satisfaction. To further explain the relation between human behavior and physical built environment regarding sense of community, sense of belonging, and sense of place, the study applies its findings to a selected study area in the city of Alexandria, Egypt. The results revealed strong correlation between physical built environment and human behavior.

Keywords: human behavior, built environment, residential satisfaction, sense of community, sense of belonging, sense of place.

1 INTRODUCTION

Many communities suffer from a lack of positive communication between individuals and the place they belong to, as well as a lack of happiness. This issue arises when urban designers disregard the role of humanity in the design process and its impact on the development of urban environments. The complexities of the place can be managed through urban design and a general framework for change can be created by designing a compatible and sustainable space for users based on the events and activities that occur in it. This shows political, social, and economic transformations that occur in communities and affect individual social-development. Despite the social differences among people of the same community, it was discovered that their behaviors toward some situations are similar, which expresses their culture [1]. The degree to which a community fits a person's needs and goals is referred to as residential satisfaction [2]. The extent to which these needs are met is determined by a person's assessment of the physical, social, and personal elements provided by their community. This includes high-quality physical-environment, community services, quality of housing, interactive networks, sense of ownership and acceptance, standards and value systems [3]. Residential satisfaction is essential as community unhappiness can affect a person's psychological well-being and quality of life [4], in addition to affecting their decision to abandon the place [5]. A variety of elements that contribute to residential satisfaction have been discovered in previous studies. The importance of the physical built environment such as parks, amenities, and housing has been highlighted in multiple

WIT Transactions on The Built Environment, Vol 210, © 2022 WIT Press
www.witpress.com, ISSN 1743-3509 (on-line)
doi:10.2495/ARC220071

researches [4]. Others have discovered social elements like belonging and social support [6] or personal considerations such as duration or residence and ownership of the home [7]. A comprehensive literature research discovered a wider range of elements that affect residential satisfaction, sense of community, sense of belonging and sense of place.

2 LITERATURE REVIEW

Literature aims at understanding relations between people and their surroundings [8]. Research shows a connection between sense of community, sense of belonging, and a sense of place through examining various ways of people's perspective towards the community, foster well-being, improve coping, form support networks, and encourage community growth [9]. The topic of residential satisfaction with its literary aspects is therefore examined and in particular the human behavior regarding sense of community, sense of belonging, and a sense of place.

2.1 Residential satisfaction

Residential satisfaction is a belief that the community satisfies personal goals and needs, and how connectivity of content and sense of community is dependent on the environment. It is a multi-dimensional system that focuses on social conditions such as membership, acceptability and the physical environment, for instance the availability of community services and the quality of housing [10]. Residents' views of the commercial, social and public services, the physical environment, the availability of opportunities and the responsibility, efficiency, and accessibility of local political decision-makers are therefore taken into account [11]. Satisfaction has been connected throughout literature with goodness and the quality of life [12]. An attempt to categorize content theories proposes two categories; content theories and process theories [12]. Content theories identified specific driving factors and needs which are favorable for satisfaction including: theory of accomplishment [13], hierarchy of necessities [14], internal–external control [15], the theory of cognitive evaluation [16], and theory of two factors [17]. On the other hand, process theories focus on dynamic thinking processes and the ways that specific behaviors and attitudes produce, including theory of drive [18]. Feelings of belongingness, community attachment, community participation, minimal fear of crime, community layout and design and housing density contribute to the experience of high levels of sense of community, sense of belonging and sense of place that lead to residential satisfaction as shown in Fig. 1.

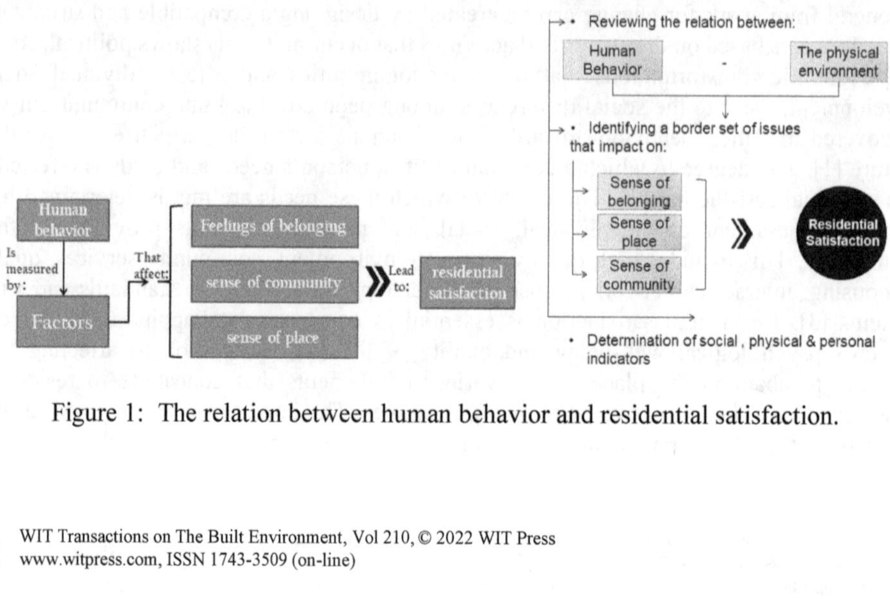

Figure 1: The relation between human behavior and residential satisfaction.

2.1.1 Sense of community

Persons with strong sense of the community believe that they can have some control over the community and be impacted by it, considering that their needs can be satisfied and met [13]. In the development of a community sense, physical, social and personal aspects have a mutual influence. Research has shown disagreement about the question of the sense of community is or has different aspects, a single, well defined concept. Because of the discussion about the basic aspects of the sense of community, a separate structure has been investigated as one of these aspects, the sense of belonging, in an attempt to understand its contribution to the sense of society. Multiple domains are mentioned in previous researches to measure the sense of community [19] presented in Table 1.

Table 1: Sense of community dimensions and it's domains [19].

Sense of community: Theoretical dimensions				
Domains of sense of community				
	Community attachment	**Community identity**	**Social interaction**	**Pedestrianism**
Primary action	Bonding with community	Identifying (with) community	Being involved in community	Knowing community
Subcomponents	Community satisfaction; Connectedness; Sense of ownership; Long-term local integration	Uniqueness; Continuity; Significance; Congruence; Cohesiveness	Neighboring; Casual social encounter; Community participation; Social support	Walkability; Pedestrian propinquity; Mass transit; Pedestrian scale/ street-level; Activities

2.1.2 Sense of belonging

People with a high feeling of membership have greater social and psychological performance, as membership offers them a sense of purpose, meaning and value [20]. The development of sense of belonging has been demonstrated by the social, physical and personal variables of participation in the community, in addition to crime fear, community layout, design, duration and marriage status [10]. When individuals feel that they belong to a community, they get more involved in it and create deep ties with others [10]. In such scenario, they feel less isolated, which leads to higher residential satisfaction as they feel that they are part of a group. There is a set of parameters in sense of belonging to place [21] as shown in Table 2.

2.1.3 Sense of place

The place is significant when it promotes a sense of goodness that unites community members and enables them to return [22]. Sense of place is a multi-dimensional structure consisting of components that increase residential satisfaction from dependency, identity and connection to each other [9]. The evolvement of feeling of place is associated with social, physical and personal elements such as community involvement, community structure, design, low crime rates and the duration of residence. The sense of place has three main domains [23] as shown in Table 3.

Table 2: The parameters in sense of belonging [21].

The parameters in sense of belonging to place				
Factors of sense of belonging				
Cultural, social and individual factors	**Physical factors**	**Background factors**	**Factor of time**	**Activity and interactive factors**
The rate of employment for social networks and cultural interaction: people's participation.	Type of access to every unit of the complex	Location position in urban areas	The length of time people is familiar with a place	Activities and interactions between humans–places
How the person is present in the place along with others.	Satisfaction from the place	How to communicate with the periphery	The rate of its use and length of stay	Activities and interactions between humans–humans
Duration of stay	Type of organization			
Life style	Geometric composition			
How to select a place	Availability of facilities			
The type of person's relationship with the place.				
Demography (age, gender, literacy)				

Table 3: Sense of place domains [23].

Sense of place domains		
Place identity	**Place attachment**	**Place dependence**
Experiences related with physical environment; Memories; Ideas; Feelings; Attitudes; Values; Preferences; Meanings and concept of behavior	Refers to the bond between people and the environment; It contains the emotional context; It is described as the affective emotional structure.	The strength of association between a person and objects; The space's ability to fulfill the needs.

3 METHODOLOGY

The study adopts a framework that is divided into several parts that aim to studying the relation between residential satisfaction and sense of community, sense of belonging, and sense of place. Firstly, it reviews previous research and literature to define the appropriate selection factors for each dimension [19], [21]. Secondly, it adopts an international urban planning guidelines [24]. Lastly, a close-ended questionnaire is designed to examine the relation between dependent variables (sense of community, sense of belonging, and sense of place) and some independent variables (built environment) in a neighborhood in Alexandria City. The selected neighborhood was relocated from an unplanned residential area that called Ma'wa El Saidin to another planned residential area with different quality of the built environment (physical attributes).

3.1 Methods and tools

Data collection: The dimensions, criteria and factors for evaluating sense of community, sense of belonging, and sense of place are selected after a thorough study of existing literature research and relevant examples. Several previous researches and studies have been invested in measuring the three senses.

3.1.1 Measuring the dependent variables

- Sense of community: Kim and Kaplan studied the physical relationship between sense of community and urban form shown in a framework for the physical and social dimensions which defines four contributing domains: (1) community attachment; (2) social interaction; (3) community identity; and (4) pedestrianism [19].
- Sense of belonging: Maleki et al. defined five main factors affecting sense of belonging: (1) cultural; (2) social and individual factors; (3) physical factors; (4) background factors; and (5) factor of time and activity and interactive factors [21].
- Sense of place: Wahyudie et al. 2021 defined three domains of sense of place: (1) place attachment; (2) place dependence; and (3) place identity [23].

3.1.2 Measuring the independent variable

The built environment is measured through an observation table of the physical characteristics of the built environment that is composed of physical components proposed by Kim and Kaplan theory, as well as built environment variables (transport, safety and danger, privacy and crowding, participation and empowerment) [25].

3.2 Statistical analysis of data

The collected data are analyzed using IBM SPSS software package version 20.0. (Armonk, NY: IBM Corp.). Qualitative data were described using number and percent. The Kolmogorov–Smirnov test is used to verify the normality of distribution. Quantitative data is described using range mean, standard deviation. Significance of the obtained results is judged at the 5% level.

The implemented tests are:

1. Student t-test, for normally distributed quantitative variables, to compare between two studied groups.
2. Pearson coefficient, to correlate between two normally distributed quantitative variables.
3. Regression, to detect the most independent/ affecting factor for affecting human behavior.

3.3 Study area

Alexandria is located in the north of Egypt and it extends from Abu-Quir Bay in the east to Borg-El-Arab in the west. It is a 90 km-long seaside city that overlooks the Mediterranean Sea. Alexandria City can be divided into the following sections: Alexandria, Borg El-Arab City, and Centre, and also New El-Arab. Alexandria is divided into six districts: Al Montaza, Sharq (east), Al-Gomrok, Al-Amreya, Agamy and Gharb (west). This study focuses on a resettlement neighborhood in the west district of Alexandria City (Fig. 2).

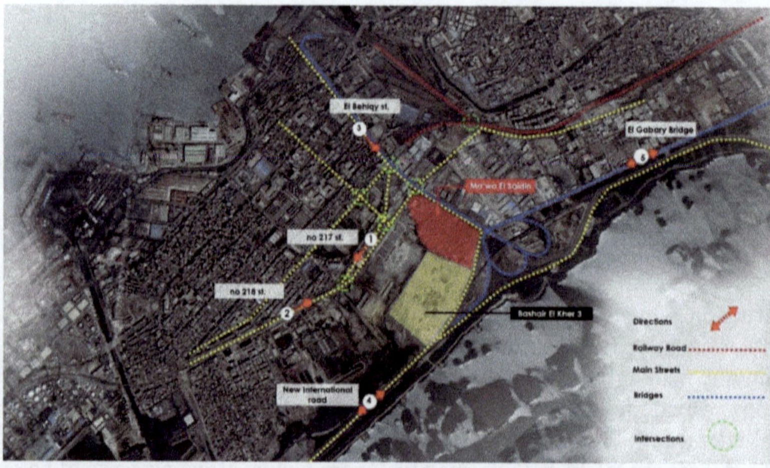

Figure 2: Study area geographic location, boundaries and accessibility.

These squatters formed as a result of the presence of navigators of King Mariout near them. In 1960, groups of fishermen settled in this area, built their houses there and they called it "Ma'awa El Saidin". The government built houses for these groups, however each family made an extension to their houses which resulted in narrowing the streets. In 1964, some individuals started filling in the lake to build more houses which resulted in the diminution of the lake from 106,000 acres to 17,000 acres. Moreover, they got rid of the sewage and the industrial waste in the lake which consequently caused its pollution and the death of fish. As a result, fishermen switched to collecting rubbish and burning it which caused air pollution and the spread of lots of diseases. Nowadays, the government relocated them to another area called "Bashair El-Kher 3" with different built environment, facilities and services.

4 RESULTS AND DISCUSSION

In this section, a group of factors is selected from the main dimensions as mentioned in the literature review (Section 2.1) to measure the effect of built environment on human behavior. An online questionnaire is applied to compare the perspective of academic personnel and the residents of "Bashayer El-Khair 3" after their resettlement from "Ma'wa El Sayden" neighborhood. The following results illustrate the online questionnaire which is classified into four main parts. The respondents of the questionnaire include 54 academic personnel and 60 residents of the selected neighborhood. The demographic data of the respondents is described in details in Table 4.

Table 4: The percentage of the responses and their demographic data.

	Academic (n = 54)		Residents (n = 60)	
	No.	%	No.	%
Gender				
Male	17	31.5	36	60.0
Female	37	68.5	24	40.0
Age/Academic				
23–35 years old	29	53.7	–	–
35–45 years old	18	33.3	–	–
45–55 years old	2	3.7	–	–
55–65 years old	5	9.3	–	–
Age/Residences				
0–15 years old	4	6.7	–	–
15–30 years old	35	58.3	–	–
30–45 years old	15	25.0	–	–
45+ years old	6	10.0	–	–
Academic degree				
Demonstrator	21	38.9	–	–
Lecturer	16	29.6	–	–
Assistant Professor	13	24.1	–	–
Professor	4	7.4	–	–
Marital status				
Single never married	16	29.6	16	26.7
Married	38	70.4	33	55.0
Divorced or widow	0	0.0	11	18.3
Education/Residences				
Primary	–	–	19	31.7
Preparatory	–	–	10	16.7
Secondary	–	–	21	35.0
University	–	–	10	16.7
Income/Residences				
<500	–	–	12	20.0
500–1000 L.E	–	–	6	10.0
1000–2000 L.E	–	–	14	23.3
2000–3000 L.E	–	–	15	25.0
3000–4000 L.E	–	–	6	10.0
4000–5000 L.E	–	–	7	11.7

The following section illustrates the degree of impact of human behavior and physical environment on the selected dimensions with comparison to the respondent's answers as shown in Table 5. If $p \leq 0.05$, then there is a significant difference among the academic and residents' perspective. The statistical analysis shows a significant difference in the impact of human behavior regarding sense of community while the physical built environment doesn't show any significant difference. Following this, the physical built environment and human behavior towards sense of place shows a significant difference. Finally, the statistics clarify that the human behavior regarding the sense of belonging doesn't show any significant difference.

Table 5: Comparison between Academic and residents perspective.

	Academic (n = 54)	Residents (n = 60)	T	P
Sense of community				
Physical environment				
Mean score	3.04±0.56	3.15±0.61	0.980	0.329
Percent score	60.86±11.17	62.98±12.11		
Human behavior				
Mean score	3.59±0.71	3.26±0.83	2.298*	0.023*
Percent score	71.85±14.28	65.14±16.62		
Sense of place				
Physical environment				
Mean score	3.07±0.91	3.47±0.96	2.230*	0.028*
Percent score	61.48±18.16	69.33±19.30		
Human behavior				
Mean score	2.81±0.64	3.23±0.82	3.032*	0.003*
Percent score	56.30±12.72	64.67±16.31		
Sense of belonging				
Human behavior				
Mean score	2.99±0.43	2.88±0.59	1.072	0.286
Percent score	59.75±8.58	57.70±11.74		

t: Student t-test; *: Statistically significant at $p \leq 0.05$

In Table 6, the statistical analysis illustrates the correlation between human behavior and the physical built environment. If the Pearson coefficient (r) is positive, then the relation between quantitative variables is direct. Therefore, the physical built environment and the human behavior show a positive significant correlation.

Table 6: Correlation between physical environment and human behavior.

	Human behavior			
	Academic (n = 54)		Residences (n = 60)	
	R	p	R	P
Physical environment/Sense of community	0.603*	<0.001*	0.517*	<0.001*
Physical environment/Sense of place	0.295*	0.030*	0.658*	<0.001*
Overall physical environment	0.439*	0.001*	0.678*	<0.001*

r: Pearson coefficient; *: Statistically significant at $p \leq 0.05$

Table 7 shows a multivariate analysis regression to clarify the most effective variable on the human behavior. From the academic perspective, the physical built environment of the sense of community has the most significant regression toward the human behavior. On the contrary, the residents' physical environment of sense of place is considered the most significant regression one. According to the results of R^2 coefficient, the previous analysis could be achieved by a prediction of (0.419) for the academic and (0.465) for the neighborhood residents'. Finally, the model is validated according to significance of f p values.

Table 7: Multivariate analysis linear regression for human behavior.

	Academic (n = 54)				Residents (n = 60)			
	B	**Beta**	**t**	**p**	**B**	**Beta**	**T**	**p**
Constant	31.424		5.912*	<0.001*	40.496		10.814*	<0.001*
Physical environment/ Sense of community	0.687	0.868	5.399*	<0.001*	0.128	0.215	1.839	
Physical environment/ Sense of place	–0.172	–0.354	2.200*	0.032*	0.201	0.537	4.589*	<0.001*
R^2	0.419				0.465			
F(p)	18.398* (<0.001*)				24.760* (<0.001*)			

F,p: f and p values for the model; R^2: Coefficient of determination; B: Unstandardized Coefficients; Beta: Standardized Coefficients; t: t-test of significance; *: Statistically significant at $p \leq 0.05$

Table 8 shows the overall physical built environment multivariate analysis regression to clarify that the P is a significant value in the academic survey and in the residents survey. Also, the results of R^2 coefficient are (0.193) for the academic and (0.460) for the residence, which mean that the model is valid.

Table 8: Multivariate linear regression for human behavior.

	Academic (n = 54)				Residents(n = 60)			
	B	**Beta**	**t**	**p**	**B**	**Beta**	**t**	**p**
Constant	45.377		9.053*	<0.001*	39.339		11.685*	<0.001*
Overall physical environment	0.282	0.439	3.527*	0.001*	0.350	0.678	7.030*	<0.001*
R^2	0.193				0.460			
F(p)	12.440* (<0.001*)				49.415* (<0.001*)			

F,p: f and p values for the model; R^2: Coefficient of determination; B: Unstandardized Coefficients; Beta: Standardized Coefficients; t: t-test of significance; *: Statistically significant at $p \leq 0.05$.

In Table 9 the observation shows that "Bashayer El-Khair" neighborhood has a total score 61% of quality of physical built environment attributes. The observation of the neighborhood revealed that its main problems are the absence of aesthetics pleasantness, artistic details and local and unique characteristics. Also, it has a bad relationship between the void and solid, high rise buildings do not respect the human scale.

5 CONCLUSION

Through the investigation of literature and relevant cases, the research focused on adopting a framework to highlight the main domains and factors of the dependent variables (sense of community, sense of belonging and sense of place) as a means to measure the human behavior and the independent variable (the physical built environment). In addition to, clarifying the relation between these variables and the residential satisfaction.

A comparative analysis survey was applied among the academic and the residents' perspective. The variables were addressed by a series of questions to measure the relation

Table 9: Observation results. *(Source: Researcher adopted from Ashraf et al. [25].)*

The built environment factors		Score
Aesthetics	The presence of major attractions in the community layout and design	2
	Aesthetics pleasantness	1
	Encourage the participation of artist in detailing of buildings	1
	Harmony	2
	The architectural style is well connected with the past	1
	The presence of local and unique characteristics	1
Total 18		8
Percentage		44%
Street and services	Mixed use neighborhood	3
	Community services	3
	Accessibility	3
	Local parking facilities	2
	Public transportation	2
	Walkable streets with good sign system which encourage pedestrian activities	2
	The presence of squares, parks, services, shops, recreation, facilities and play areas in their right place	2
Total 21		17
Percentage		80%
Buildings	Identifying population density	3
	Length of residence	2
	Building lines respecting, window proportions and solid to void relationships	1
	The presence of affordable housing	2
	The respect of building frontage that provides encloses and definition to streets and spaces	2
	Sense of human scale in high rise buildings	1
	The lower floor should be differentiated architecturally	1
Total 21		12
Percentage		57%
Total 60		37
Percentage		61%

Likert scale (10): Bad or not exist = 1, Fair = 2, Good = 3.

between the human behavior and the physical built environment regarding sense of community, sense of belonging and sense of place in a selected neighborhood in Alexandria City.

The future research should aim to investigate the concluded variables within different neighborhoods in Alexandria City to compare the responses towards the same variables, which could indicate the degree of impact of different urban patterns on the human behavior.

REFERENCES

[1] Moussa, R.A., The impact of societal transformations on the urban spaces during the local festivals. *Acad. Res. Community Publ.*, **2**, p. 183, 2018. DOI: 10.21625/archive.v2i3.358.

[2] Max, L., Determinants of residential satisfaction: Ordered logit vs. regression models. *Growth Change*, **30**, pp. 264–287, 1999. DOI: 10.1111/0017-4815.00113.

[3] Smith, K.M., *Edith Cowan University: Grants Register 2022*, pp. 376–377, 2021. DOI: 10.1057/978-1-349-96042-2_425.

[4] Braubach, M., Residential conditions and their impact on residential environment satisfaction and health: Results of the WHO large analysis and review of European housing and health status (LARES) study. *Int. J. Environ. Pollut.*, **30**, pp. 384–403, 2007. DOI: 10.1504/IJEP.2007.014817.

[5] James, R.N., Sweaney, A.T. & Sweaney, A.L., Sources of discontent: Residential satisfaction of tenants from an internet ratings site. *Environ. Behav.*, **41**, pp. 43–59, 2009. DOI: 10.1177/0013916507310031.

[6] Adriaanse, C.C.M., Measuring residential satisfaction: A residential environmental satisfaction scale (RESS). *J. Hous. Built Environ.*, **22**, pp. 287–304, 2007. DOI: 10.1007/s10901-007-9082-9.

[7] Campbell, C., Foucault, psychology and the analytics of power by Hook, Derek. Basingstoke: Palgrave MacMillan, 2007 (book review). *Journal of Community and Applied Social Psychology*, **18**(6), pp. 645–646, 2008. DOI: 10.1002/casp.994.

[8] Bonaiuto, M., Aiello, A., Perugini, M., Bonnes, M. & Ercolani, A.P., Multidimensional perception of residential environment quality and neighborhood attachment in the urban environment. *J. Environ. Psychol.*, **19**, pp. 331–352, 1999. DOI: 10.1006/jevp.1999.0138.

[9] Altman, I. & Low, S.M., Place Attachment, vol. 148, 1992.

[10] Grillo, M.C., Teixeira, M.A. & Wilson, D.C., Residential satisfaction and civic engagement: Understanding the causes of community participation. *Soc. Indic. Res.*, **97**, pp. 451–466, 2010. DOI: 10.1007/s11205-009-9511-0.

[11] Puddifoot, J.E., Community identity and sense of belonging in a northeastern English town. *J. Soc. Psychol.*, **134**, pp. 601–608, 1994. DOI: 10.1080/00224545.1994.9922990.

[12] Allen, L.R., Long, P.T. & Perdue, R.R., Relational patterns between community dimensions and global measures of community satisfaction. *J. Rural Stud.*, **7**, pp. 331–338, 1991. DOI: 10.1016/0743-0167(91)90095-A.

[13] McMillan, D.W. & Chavis, D.M., Sense of community: A definition and theory. *J. Community Psychol.*, **14**, pp. 6–23, 1986. DOI: 10.1002/1520-6629(198601)14:1<6::AID-JCOP2290140103>3.0.CO;2-I.

[14] Thorpe, L.P. & Schmuller, A.M., Motivation and personality. *Personal An Interdiscip Approach*, pp. 41–65, 2010. DOI: 10.1037/12225-003.

[15] Rotter, J.B., Generalized expectancies for internal versus external control of reinforcement. *Psychological Monographs: General and Applied*, **80**(1), pp. 1–28, 1966.

[16] Deci, E.L., Nezlek, J. & Sheinman, L., Characteristics of the rewarder and intrinsic motivation of the rewardee. *J. Pers. Soc. Psychol.*, **40**, pp. 1–10, 1981. DOI: 10.1037/0022-3514.40.1.1.

[17] Steers, R.M. & Mowday, R.T., Introductin to special topic forum: The future of work motivation theory. *The Academy of Management Review*, **29**(3), pp. 379–387, 2004.

[18] Ryan, R.M. & Deci, E.L., Intrinsic and extrinsic motivations: Classic definitions and new directions. *Contemporary Educational Psychology*, **25**(1), pp. 54–67, 2000. DOI: 10.1006/ceps.1999.1020.
[19] Kim, J. & Kaplan, R., Physical and psychological factors in sense of community: New urbanist Kentlands and nearby orchard village. *Environ. Behav.*, **36**, pp. 313–340, 2004. DOI: 10.1177/0013916503260236.
[20] Newman, B.M., Lohman, B.J. & Newman, P.R., Peer group membership and a sense of belonging: Their relationship to adolescent behavior problems. *Adolescence*, **42**(166), 241–263, 2007.
[21] Maleki, B., Casanovas Rubio, M.D.M., Hosseini, S.M.A., De La Fuente Antequera, A., Multi-criteria decision making in the social sustainability assessment of high-rise residential buildings. *IOP Conf. Ser. Earth Environ. Sci.*, **290**, 2019. DOI: 10.1088/1755-1315/290/1/012054.
[22] Tonts, M. & Atherley, K., Competitive sport and the construction of place identity in rural Australia. *Sport Soc.*, **13**, pp. 381–398, 2010. DOI: 10.1080/17430431003587947.
[23] Wahyudie, P., Antariksa, A., Wulandari, L.D. & Santosa, H., Place attachment in supporting the preservation of religious historical built environment. *IOP Conf. Ser. Earth Environ. Sci.*, **737**, 2021. DOI: 10.1088/1755-1315/737/1/012035.
[24] Toggia, P., International guidelines on victimization. *Encycl. Criminol. Crim. Justice*, pp. 1–7, 2013. DOI: 10.1002/9781118517383.wbeccj098.
[25] Ashraf, P., Ayad, H. & Saadallah, D., Sense of community and built environment: How can built environment, social economic conditions and history of place shape our sense of community? *Int. J. Eng. Adv. Technol.*, pp. 2249–8958, 2017.

SERVICE DESIGN AND QUALITY: RESTAURANTS IN BAHRAIN

FATEMA SEDDIQ, KHADIJA HASSOUNA, HANA AL JOWDER & FATEMA QAED
Department of Architecture and Interior Design, College of Engineering, University of Bahrain, Bahrain

ABSTRACT

The purpose of this paper is to investigate service design and quality of restaurants in Bahrain and an understanding of its importance in the growing restaurant industry in Bahrain. This paper investigates the service design and restaurant design of restaurants in the Seef District, Bahrain using the DINESERV scale. A questionnaire was used to rate seven restaurants in the Seef District and the ratings are tabulated and a total mean out of 7 is calculated. The ratings for all of the seven restaurants were 5 and higher with minor differences between the participants ratings. This signifies that the level of service design and restaurant design quality in the Seef District is relatively high which subsequently leads to high customer satisfaction and high visiting rates. This brings to attention that there is a high level of awareness of the importance of service design and restaurant design in that area. Moreover, the final data brought to attention the gender, age and working status of the visiting participants. The data showed that most of the visitors were mid-20s, female students.

Keywords: service design, tools, restaurant design, service quality.

1 INTRODUCTION

The concept of service design has been utilized for decades before its realization to describe a service and its delivery [1]. Service design is a term that relates to a collection of physical and non-physical goods that involve multidisciplinary design in order for customers to be able to access and utilize them successfully and delightfully [2]. In the field of restaurants' service design, many elements impact the efficiency and the experience of restaurant quality like food quality, the physical atmosphere and restaurant's design, and the most important is the services quality [3], [4]. The services provided in the restaurants and the way they are designed is what pushes the consumer to revisit the place and develop a connection with it.

Bahrain, according to a 2013 Economy Freedom Index, has an increased demand for dining out and is considered an ideal spot in the Middle East to invest in the restaurant industry. Despite this, there is insufficient data about its restaurants' design and service design quality. This paper investigates the service design and quality of restaurants in Bahrain using the DINESERV scale.

2 SERVICE DESIGN AND RESTAURANT DESIGN

2.1 Service design in restaurants

With vast varieties of restaurants to serve 7.9 billion people around the world many restaurants are in competition with each other to provide the best to please and win customers which is a vital aspect for the business's success [5]. To stand out from the rest of the restaurant businesses restaurants have opted different service design-based thoughts that deal with combining both tangible and intangible aspects to provide a complete experience for the user and attract customers, this gave rise to the notion of service quality in restaurants [6].

Some researchers define service qualities in restaurants as the actions that can be translated to values [7]. Conversely, according to Grönroos [8], service quality is considered similar to customer's satisfaction, which is the perception of the customer of the service

WIT Transactions on The Built Environment, Vol 210, © 2022 WIT Press
www.witpress.com, ISSN 1743-3509 (on-line)
doi:10.2495/ARC220081

provided in a restaurant when compared to his expectation. In an attempt to better understand the notion, Grönroos [9] and Lehtinen and Lehtinen [10] agreed on identifying three aspects of service quality which are the technical quality of the result like products/services, the functional quality of the experience which is considered as the interaction between the provider and the consumer, and the business company identity. Sasser et al. [11] had different views about the idea they came up with seven qualities that sufficiently cover the idea of service quality which include: Security, attitude, completeness, condition, consistency, availability, and training.

As stated above various researchers had various definitions for the word service quality which makes it difficult to define [12]. These issues stem mostly from the way services were created, used, and assessed thus, service quality, according to Grönroos [8] and Parasuraman et al. [12], is more challenging than quality of product. The physical characteristics of goods, problems in production can be quickly detected, and alternatives can be implemented right away to fix the flaws before they are sold to customers. Service quality is even harder to measure physically when compared to product quality since services are abstract things with qualities such as intangible, impermanent, diverse, and instantaneous [13].

To assess service quality, Parasuraman et al. [12] developed a 22-item SERVQUAL scale which stands for (SERVICE QUALITY) with five dimensions: tangible which includes physical aspects and materials, reliable which includes efficiency consistency, ability to respond which is the readiness of the employees to assist the customers, assurance which is the employee's knowledge in dealing with the customers, and empathy which relates to the personalized, attentive care that each consumer receives. Service quality, according to the SERVQUAL instrument, it is defined as the gap between customers' expectations and their presumed impressions of the actual service, also known as the expectations method. Customers' expectations and views are measured after service is given using this technique.

A modified version of SERVQUAL was created in response to the critiques [14]. Cronin and Taylor [15] suggest SERVPERF (SERVICE PERFORMANCE), an efficiency gap assessment instrument that eliminates the expectation part of the SERVQUAL model. In terms of an overall measure of service quality, the SERVPERF instrument outperforms the SERVQUAL scale [15]. SERVPERF is able to explain most of the variation in an overall measure of service quality across different types of restaurant industries: fast food, causal and fancy when compared to the SERVQUAL scale [15]. Moreover, SERVPERF is has proven to be more valid and reliable [12].

A third scale was later developed by Stevens et al. [16], because SERVQUAL also proved to be an inappropriate model for the specific context of restaurant services. DINESERV (DINE SERVICE) was put to the test in three restaurant divisions fine, casual, and fast service dining, and it proved to be quite reliable. Although SERVQUAL, SERVPREF, and DINESERV have been utilized in the food sector, they have limitations where they do not include food quality as it relates to the product itself, which has shown to be an essential dimension in the food industry [17].

2.2 Restaurant design

Part of the service design for a restaurant includes the restaurant's design, which hugely impacts the customer's satisfaction and the quality of the services provided by the restaurant [18]. Many academics have agreed that the restaurant design provides certain ambience to the customer that deals with how they perceive their surrounding environment. This in return affects the customer's satisfaction level, recommendation to others and their rate of visiting.

The elements that create the ambience include: ventilation, colour, lighting, exterior/interior layout and music [19]–[22].

The use of adequate colours in a restaurant's design may make the space functional and appealing to customers, which can have an influence on visitor numbers [23]. While deep, warm, and informal light tones seem to promote relaxation, bright colours typically attract and provoke emotions [24]. Red, orange, and yellow are hot colours, whereas green, blue, and violet are cool hues. These colours are divided into warm and pale hues. People can see the impression of pleasant, comfortable settings through warm colours [25]. Additionally, cool hues create welcoming spaces that increase consumer satisfaction [26].

Many researchers agreed that light affects the customer emotions [27]. The consumer needs proper lighting in a restaurant so they can see the cuisine at its finest. As a result, the restaurant needed to manually illuminate each table in order to provide its unique ambiance [28]. For the purpose of enjoying the services and experience, light may create the environment cool and tranquil. When the lighting is dark, customers talk more politely [29]. One of the physical elements with the most effect on restaurants might be light [30]. A restaurant owner also wishes to convey a sense of light [31]. The location of light will also provide the customer with a unique experience.

3 RESEARCH QUESTION

According to a 2013 Economy Freedom Index, Bahrain has a restaurant business in line with increased demand for dining out and is considered as one of the perfect spots in the Middle East to invest in the restaurant industry, yet there is insufficient data about the restaurants' design and service design quality which include the physical and non-physical customer experience in Bahrain. This paper investigates the service design and quality of restaurants in Bahrain using the DINESERV scale.

4 RESEARCH METHODOLOGY

This research is descriptive research as it uses a structured process (the DINESERV scale), and the main research methodology used was the quantitative methodology, as the DINESERV scale uses a questionnaire with scalable answers. There are seven different restaurants in the Seef District, Bahrain that were evaluated for their service design and restaurant design using the DINESERV scale. The seven restaurants chosen are Le Chocolat, Café Lilou, Paul Bakery and Restaurant, Mado Café and Restaurant, Nino, Nu Asia, and Fatto. The restaurants chosen have been limited to one region in Bahrain (Seef District) to maintain a fair comparison, and the Seef District was chosen because of the high density of restaurants present in the area.

Out of the three scales mentioned in the literature review, the DINESERV scale has been used to evaluate the service design of the restaurants because of its reliability and validity. Since the SERVQUAL scale has been criticized for criterion related validity issues, overlap among the different dimensions, and unstable dimensionality, the DINESERV was chosen [30], [31]. The DINESERV scale works by considering 29 statements in a questionnaire which are indicative of a restaurant with a good service design and quality, and each statement is rated from a scale of one (strongly disagree) to seven (strongly agree). Statements 1–10 are related to tangibles, statements 11–15 are related to reliability, statements 16–18 are related to responsiveness, statements 19–24 are related to assurance, and statements 25–29 are related to empathy. The 29 statements are as follows:

1. Visually attractive parking areas and building exteriors
2. Visually attractive dining area
3. Clean, neat, and appropriately dressed staff

4. Restaurant's décor typical of its image and price range
5. Easily readable menu
6. Visually attractive menu that reflects the restaurant's image
7. Comfortable and easy to move around in dining room
8. Clean rest rooms
9. Clean dining areas
10. Comfortable seats in the dining room
11. Service in the promised time
12. Quick correction of wrong service
13. Dependable and consistent restaurant
14. Accurate bill
15. Error-free served order
16. Maintaining speed and quality of service during busy times
17. Provision of prompt service
18. Extra effort for handling special requests
19. Employees can answer questions completely
20. Comfortable and confident in dealings with the restaurant
21. Staff provide information about menu items, their ingredients, and methods of preparation
22. Feeling personally safe
23. Staff are well trained, competent, and experienced
24. Restaurant supports the employees so that they can do their jobs well
25. Employees are sensitive to individual needs and wants, rather than always relying on policies and procedures
26. Makes customers feel special
27. Anticipates customers' individual needs and wants
28. Employees are sympathetic and reassuring if something is wrong
29. Having the customers' best interests at heart

In the marketing field, the DINESERV questionnaire is a gap theory model and it compares a service quality expectation index to a service quality perception index. However, for this research the focus was on the existing surface quality perception by the customers so the gap theory model was not used. The questionnaire was carried out online using Google Forms, and the population for this study are individuals residing in Bahrain over the age of 18 who have visited one or more of the aforementioned seven restaurants.

The sampling technique used for this research was the voluntary response sample as the questionnaire link is available online and individuals can choose whether to participate or not. The strength of this technique is that it can be shared easily amongst participants as it is widely available online. The weaknesses of this technique are that it is limited to people who have the link, which can be biased towards certain gender or age group. This sampling technique was chosen as it is the easiest to share and get participants' responses within a short span of time.

The questionnaire was created on 28 April 2022 and gathered 243 responses. Considering that the population of individuals over 18 in Bahrain is over 1,000,000 according to the Ministry of Information Affairs, carrying out the research with a confidence level of 95% and obtaining 243 responses means that the margin of error is 6%. The answers from the questionnaire were entered into the Microsoft Excel program in order to get the mean rating out of 7 for each of the 29 statements for each restaurant.

5 DATA COLLECTION

The total mean value of the ratings is calculated separately for each of the seven restaurants, as well as the percentages related to the variables of the population.

Table 1: Ratings of the restaurants in the Seef District.

Le Chocolat	
Dimension	Total mean rating out of 7
Tangibles	5
Reliability	5
Responsiveness	5
Assurance	5
Empathy	5
La Chocolat total mean rating out of 7	5
Café Lilou	
Dimension	Total mean rating out of 7
Tangibles	6
Reliability	6
Responsiveness	6
Assurance	6
Empathy	6
Café Lilou total mean rating out of 7	6
Paul Bakery and Restaurant	
Dimension	Total mean rating out of 7
Tangibles	6
Reliability	6
Responsiveness	6
Assurance	6
Empathy	6
Paul Bakery and Restaurant total mean rating out of 7	6
Mado Café and Restaurant	
Dimension	Total mean rating out of 7
Tangibles	6
Reliability	5
Responsiveness	6
Assurance	6
Empathy	6
Mado Café and Restaurant total mean rating out of 7	6
Nino	
Dimension	Total mean rating out of 7
Tangibles	6
Reliability	6
Responsiveness	6
Assurance	6
Empathy	6
Nino total mean rating out of 7	6

Table 1: Continued.

Nu Asia	
Dimension	Total mean rating out of 7
Tangibles	5
Reliability	5
Responsiveness	6
Assurance	5
Empathy	5
Nu Asia total mean rating out of 7	5
Fatto	
Dimension	Total mean rating out of 7
Tangibles	6
Reliability	6
Responsiveness	5
Assurance	6
Empathy	6
Fatto total mean rating out of 7	6

Table 2: Variables' characteristics.

Variables	Characteristics	Percentage
Gender	Male	36%
	Female	64%
Age group	18–25	46%
	25–35	36%
	35–45	18%
	Above 45	0%
Job status	Unemployed	9%
	Student	46%
	Full employed	36%
	Part employed	9%

6 RESEARCH FINDINGS AND DISCUSSION

The DINESERV questionnaire was used to rate the service design and restaurant design of seven restaurants in the Seef District and the ratings are tabulated and a total mean was calculated. The questionnaire was carried out online using Google Forms, and the population of the study was Bahraini residents over the age of 18, who have visited one or more of the seven restaurants. The questionnaire gathered 243 responses.

6.1 Findings and discussion for the five dimensions

Based on the findings from the seven restaurants, the restaurants in the Seef District, Bahrain generally scored well based on the DINESERV scale. The restaurants got an overall rating between 5 and 6 out of 7, generally indicating satisfaction about the service design and quality of the restaurants. Out of the seven restaurants, Paul Bakery and Restaurant, Mado Café and Restaurant, Nino, and Fatto scored the highest, with a mean rating of 6 out of 7. Le Chocolat, Café Lilou, and Nu Asia scored the lowest, with a mean rating of 5 out of 7. Overall, almost all restaurants got a similar mean rating for the five dimensions, indicating that the performance is similar in these dimensions per restaurant.

6.1.1 The dimension of tangibles

The dimension of tangibles relates to the visual appearance of the restaurant's exterior and interior (including WCs), the appearance and presentation of staff, the menu design, the comfort of restaurant circulation, and comfortable furniture. Although all restaurants scored similarly in this dimension, the restaurants of Café Lilou, Paul Bakery and Restaurant, Mado Café and Restaurant, Nino, and Fatto scored slightly higher (6 out of 7), while Le Chocolat and Nu Asia scored slightly lower (5 out of 7). Overall, the restaurants scored a satisfactory rating, indicating good visual appearance of the restaurant, cleanliness and good maintenance, good menu design, and physical and visual comfort. This is likely due to the restaurants chosen leaning towards being categorized as expensive, so the presentation of the restaurant is important in creating an attractive, professional, and reliable image that reflects the price range of the restaurant.

On average, the restaurants scored the lowest in the statements related to the appearance of staff (the lowest being 4 out of 7, neutral), and highest in the statements related to the restaurant design and decor (the highest being 7 out of 7, extremely satisfied). This reflects that most restaurants put a lot of consideration while hiring interior designers and architects, possibly because the design of the restaurant can be a marketable aspect especially in the digital age. For example, customers taking photos of the interior and posting it online, indirectly promoting the restaurant; or the restaurant posting photos on its social media platforms for direct promotion.

Regarding the restaurant design in the Seef District, the restaurants use various materials depending on the atmosphere intended to be created, typically linked to the type of food served. For example, the restaurant Paul Bakery and Restaurants uses materials and decor reminiscent of 19th century restaurants in Europe (such as the plaster mouldings in the walls) due to the restaurant sticking to the same atmosphere and décor since it was founded in the late 19th century. Another example is the restaurant of Nu Asia using décor and materials reminiscent of traditional East Asian architecture (such as the patterns and style of the partitions used and the prominence of the maroon colour) as it serves food from that region. The restaurants of Le Chocolat and Café Lilou use décor and materials reminiscent of a Parisian café (such as the use of elaborate décor and attention to details in picture frames, furniture etc.) as they serve French patisserie and food.

Regardless of the variety of the atmospheres of the restaurants, the restaurant design shares some common elements. One example is the common use of natural materials, such as the common use of the wood parquet flooring in the restaurants. The most common colour scheme used is the natural colour scheme as well, with shades of brown being the most prominent. Moreover, the restaurants commonly use wall and floor finishings that are contrasting, such as the use of glossy porcelain flooring and matte gypsum walls in Le Chocolat, and the use of brick walls and wood parquet flooring in the restaurant of Fatto. In addition, most of the restaurants are located within malls or part of a larger building complex as standalone restaurants are not very ideal in Bahrain due to most areas not being very walkable due to the extreme heat of Bahrain, especially during the summer months. As a result, most of the commercial activities are located within malls and shopping complexes. Moreover, most of the restaurants utilize the outdoor areas during specific months of the year, usually during the colder months as the weather can be too hot during the summer for outdoor dining.

Light plays a crucial role in providing the users with desirable ambiances and can aid in providing them with the required atmosphere for a good dining experience. The seven restaurants shown in Fig. 1 have mixed types of lightings like: spot lights, indirect cove lights, pendent lights, chandeliers and table lamps. The use of these types of lights provided consistency in the distribution of lighting within the restaurants, it also aids in defining different spaces in the dining area. The use of chandeliers has provided aesthetic appearance that goes along with the desired interior style chosen for that particular restaurant, it also gave the space sense of luxury resembling fine dining experience. The table lamps and pendent lights in restaurants directed light towards the tables to provide more quality to the food dishes placed on the tables, this plays an important role in making the dishes look more vivid and visually appealing.

Figure 1: Restaurants in the Seef District, Bahrain. Top row, from left to right: Le Chocolat, Café Lilou, Paul Bakery and Restaurant, Mado Café and Restaurant. Bottom row, from left to right: Nino, Nu Asia, Fatto. *(Source: Mohammad, 2021; Alzahrani, 2022; Kooheji, 2022; Alzahrani, 2021; Alghasra, 2020; Yafai, 2018; Almahrous, 2022.)*

The furniture used in the restaurants are of different styles: Le Chocolat, Café Lilou, Paul Bakery and Restaurant, Mado Café and Restaurant have used classical dining individual chairs with round or square tables, the chairs have cushioned support to ensure comfortable dining experience and allow visitors to spend long hours in the place. Nu Asia restaurant have used Asian inspired furniture to provide a certain atmosphere to the customers that match the dishes provided in order to deliver a complete dining experience. On the other hand, Nino and Fatto have opted a more modern approach of using a more contemporary furniture design making the space more efficient by ensuring fitting the maximum number of visitors at the same time. The furniture used mostly is semi fixed or fixed to control the customer behaviour and maintain the space organization.

6.1.2 The dimension of reliability

The dimension of reliability relates to the accuracy, punctuality, and reliability of the restaurant in preparing and delivering its dishes. Most restaurants scored similarly in this dimension, with the restaurants of Café Lilou, Paul Bakery and Restaurant, Nino, and Fatto scoring slightly higher (6 out of 7), while Le Chocolat, Mado Café and Restaurant, and Nu Asia scoring slightly lower (5 out of 7). This indicates overall satisfaction with the reliability of the restaurant.

6.1.3 The dimension of responsiveness

The dimension of responsiveness, similarly to reliability, relates to the punctuality and good time management of the restaurant, particularly during busy periods. Even though the restaurants scored similarly in this dimension, the restaurants of Café Lilou, Paul Bakery and Restaurant, Mado Café and Restaurant, Nino, and Nu Asia scored slightly higher (6 out of 7), while Le Chocolat and Fatto scored slightly lower (5 out of 7). This indicates overall satisfaction with the punctuality and time taken to provide the restaurant's services.

Considering that the previous dimension also had a similar score (5–6 out of 7, indicating satisfaction), this shows the attention paid to the punctuality and time management of restaurants in the Seef District, Bahrain. This indicates that it is important to the customers, and could be a huge negative aspect for a lot of customers if the service is not timely and prompt.

6.1.4 The dimension of assurance

The dimension of assurance relates to the restaurant's services and staff in terms of competence, knowledgeability about the dishes and services, and experience. Although the seven restaurants all scored similarly in this dimension, the restaurants of Café Lilou, Paul Bakery and Restaurant, Mado Café and Restaurant, Nino, and Fatto scored slightly higher (6 out of 7), while Le Chocolat and Nu Asia scored slightly lower (5 out of 7). This indicates overall satisfaction with the staff's experience, competency, and knowledgeability, as well as feeling safe and confident in the restaurant's services. Considering the relatively expensive nature of these restaurants, it is important to be able to stand out and provide very adequately trained and experienced staff so the customers are confident in dealing with the restaurant.

6.1.5 The dimension of empathy

The dimension of empathy relates to how empathetic the restaurant's services and staff are towards the individual needs of the customers, and having the customers' best interests at heart. The restaurants all scored similarly in this dimension, but the restaurants of Café Lilou, Paul Bakery and Restaurant, Mado Café and Restaurant, Nino, and Fatto scored slightly higher (6 out of 7), while Le Chocolat and Nu Asia scored slightly lower (5 out of 7). This indicates overall satisfaction with how empathetic and understanding the restaurant's staff are. The dimension of empathy is an important dimension in deciding the customer's satisfaction with the restaurant, since it is more personal and helps the customers create a more personal connection with the restaurant, so scoring strongly in this dimension could help overlook some of the other flaws in the service of the restaurant.

6.2 Findings and discussion for the characteristics of the respondents

Despite efforts to balance the sample in terms of gender and age, there is a predominance of 18–25 years old female students. The sampling was biased towards this group due to being the most common recipient of the survey shared online.

6.2.1 Gender

According to the data collected, more female participants have visited the restaurants and showed more positive ratings than male participants to the listed restaurants. Women tend to be more social and require more time with friends and gatherings. Moreover, according to Ryabova [32], women spend more time in restaurants than men. Women have also shown more interest and are more attracted to the services and aesthetics of the place than the food [33]. Based on observations of the selected restaurants during visits more female visitors were present at the restaurants during various hours of the day.

6.2.2 Age and job status

Based on the gathered responses, students aged 18–25 were most likely to visit the restaurants. This behaviour of dining at restaurants for these age groups is due to social and psychological reasons. Restaurant dining has become a social symbol, it stands as a social event, according to Knutson and Patton [34], which acts as a motive for dining out. According to Lillicrap and Cousins [35], and Hitti [36], social connection is a desire to enjoy the company of professional colleagues as well as leisure, when individuals choose to dine with family and friends at a restaurant. Students and full job timers also tend to eat more in restaurants due to convenience to save time and effort in the kitchen [37].

7 CONCLUSION

The number of restaurants in Bahrain have increased to 660,755, according to the US Department of Agriculture [38]. In the most recent quarter, this statistic climbed by more than 2% over the preceding quarter. With the restaurant sector rapidly expanding, many academics are interested in studying the influence of service design on consumer satisfaction. In this research the service design of the Seef District restaurants has been investigated in order to find out the quality of service design and restaurant design, and the level of customer satisfaction. The results indicate that restaurants' design, presentation, quickness of service, competence, and empathy have the most impact.

With these findings, it is clear that the restaurants management and designers should focus on those criteria in order to strengthen efforts to achieve the maximum level of customer satisfaction to gain more loyal customers which will in turn thrive the restaurant industry [39]. This study emphasizes the importance of restaurant service design, and the outcomes of this study bring fresh knowledge to the literature about restaurants design in Bahrain from a theoretical standpoint.

ACKNOWLEDGEMENTS

Special thanks to Dr Hana Al Jowder and Dr Fatema Qaed, assistant professors at the Department of Architecture and Interior Design, College of Engineering, University of Bahrain, for supervision of this research paper. This research did not receive any grants or funding.

REFERENCES

[1] Holopainen, M., Exploring service design in the context of architecture. *The Service Industries Journal*, **30**(4), pp. 597–608, 2010.

[2] Gloppen, J., Perspectives on design leadership and design thinking and how they relate to European service industries. *Design Management Journal*, **4**(1), pp. 33–47, 2009.

[3] Dulen, J., Quality control. *Restaurant and Institutions*, **109**(5), pp. 38–52, 1999.

[4] Susskind, A.M. & Chan, E.K., How restaurant features affect check averages: A study of the Toronto restaurant market. *The Cornell Hotel and Restaurant Administration Quarterly*, **41**(6), pp. 56–63, 2000.
[5] Gilbert, G.R., Veloutsou, C., Goode, M.M. & Moutinho, L., Measuring customer satisfaction in the fast food industry: A cross-national approach. *Journal of Services Marketing*, **18**(5), pp. 371–383, 2004.
[6] Marinkovic, V., Senic, V., Ivkov, D., Dimitrovski, D. & Bjelic, M., The antecedents of satisfaction and revisit intentions for full-service restaurants. *Mark. Intell. Plan*, 2014.
[7] O'Sullivan, J., Edmond, D. & Ter Hofstede, A., What's in a service? *Distributed and Parallel Databases*, **12**(2), pp. 117–133, 2002.
[8] Grönroos, C., A service quality model and its marketing implications. *European Journal of Marketing*, 1984.
[9] Grönroos, C., Strategic management and marketing in the service sector. Student literature, Chartwell-Bratt, Sweden, 1991.
[10] Lehtinen, U. & Lehtinen, J.R., Service quality: A study of quality dimensions. Unpublished working paper, Service Management Institute: Helsinki, Finland, 1982.
[11] Sasser, W.E., Olsen, R.P. & Wyckoff, D.D., *Management of Service Operations*, Allyn & Bacon: Boston, MA, 1978.
[12] Parasuraman, A., Zeithaml, V. & Berry, L.L., Delivering quality service. *Quarterly*, **41**(6), pp. 56–63, 1990.
[13] Lovelock, C., The dot-com meltdown: What does it mean for teaching and research in services? *Managing Service Quality: An International Journal*, 2001.
[14] Ladhari, R., Alternative measures of service quality: A review. *Managing Service Quality: An International Journal*, 2008.
[15] Cronin, J.J. & Taylor, S.A., Measuring service quality: A re-examination and extension. *Journal of Marketing*, **56**(3), pp. 55–68, 1992.
[16] Stevens, P., Knutson, B. & Patton, M., DINESERV: A tool for measuring service quality in restaurants. *The Cornell Hotel and Restaurant Administration Quarterly*, **36**(2), pp. 5–60, 1995.
[17] Cao, Y., Comparison of customers' perceptions of service quality between different management forms in fast food restaurants, Doctoral dissertation, 2011.
[18] Chiguvi, D., Impact of ambience conditions on customer satisfaction in the restaurant industry: Case study of Debonairs Pizza outlets in Botswana. *International Journal of Science and Research*, **6**(2), pp. 1825–1833, 2017.
[19] Oliver, R.L., *Satisfaction: A Behavioral Perspective on the Consumer*, McGraw-Hill: New York, 1997.
[20] Jordaan, Y. & Prinsloo, M., Grasping service markets, 2001.
[21] Kivela, J., Bakaran, R. & Reece, J., Consumer research in the restaurant environment, Part 1: A conceptual model of dining satisfaction and return patronage. *International Journal of Contemporary Hospitality Management*, **11**(5), pp. 205–222, 1999.
[22] Choi, T.Y. & Chu, R., Determinants of hotel guests satisfaction and repeat patronage in the Hong Kong hotel industry. *International Journal of Hospitality Management*, **3**(20), pp. 234–244, 2001.
[23] Kurtz, D.L. & Clow, K.E., *Services Marketing*, Wiley, 1998.
[24] Stroebele, N. & De Castro, J., Effect of ambience on food intake and food choice. *Nutrition*, **20**(9), pp. 821–838, 2004.

[25] Kivela, J., Inbakaran, R. & Reece, J., Consumer research in the restaurant environment, Part 1: A conceptual model of dining satisfaction and return patronage. *International Journal of Contemporary Hospitality Management*, **11**(5), pp. 205–222, 1999.
[26] Motoki, K., Saito, T., Nouchi, R., Kawashima, M. & Sugiura, M., Light colors and comfortable warmth: Cross modal correspondences between thermal sensations and color lightness influence consumer behaviour. *Food Quality and Preference*, **72**, pp. 45–55, 2019.
[27] Sharifi, S., When more pain is better: Role of need for uniqueness on service evaluations of observable service recovery. *International Journal of Hospitality Management*, 2019.
[28] Ciani, A.E., Thesis and dissertation. Digital Repository at Lawa State University, 2014.
[29] Yuksel, A. & Yusel, F., Measurement of tourist satisfaction with restaurant services: A segment-based approach. *Journal of Vacation Marketing*, **9**(1), pp. 52–68, 2002.
[30] Steffy, G.R., *Architectural Lighting Design*, 3rd ed., Wiley: New York, 2008.
[31] Flynn, J.E., A study of subjective responses to low energy and nonuniform lighting systems. *Lighting Design and Application*, 7(2), pp. 6–15, 1977.
[32] Ryabova, T.B., Stereotypes and stereotyping as a problem of gender studies. *Personality. The Culture. Society*, **1–2**(15–16), pp. 120–139, 2003.
[33] Driskell, J.A., Meckna, B.R. & Scales, N.E., Differences exist in the eating habits of university men and women at fast-food restaurants. *Nutrition Research*, **26**(10), pp. 524–530, 2006.
[34] Knutson, B.J. & Patton, M.E., Restaurants can find gold among silver hair. *Journal of Hospitality and Leisure Marketing*, **1**(3), pp. 79–90, 1993.
[35] Lillicrap, D.R. & Cousins, J.A., *Food and Beverage Service*, 7th ed., Hodder and Stoughton: London, 2006.
[36] Hitti, M., Top 11 reasons for fast food popularity. *Journal of the American Dietetic Association*, **108**(12), pp. 2066–2070, 2008.
[37] Davis, B., Lockwood, A., Alcott, P. & Pantelidis, L., *Food and Beverage Management*, 5th ed., Routledge: London, 2012.
[38] Santorini Chicago. How many restaurants in Bahrain? https://www.santorinichicago.com/how-many-restaurants-in-bahrain/. Accessed on: 3 May 2022.
[39] Skogland, I. & Siguaw, J.A., Are your satisfied customers loyal? *Cornell Hotel and Restaurant Administration Quarterly*, **45**(3), pp. 221–234, 2004.

SECTION 4
SUSTAINABLE CITIES

HOW TO BUILD FOOD-SUSTAINABLE CITIES AND GIVE HEALTH TO THE AGEING URBAN RESIDENTS: A TALE OF TWO COMMUNITY GARDENS IN LISBON, PORTUGAL

ISABEL MARIA MADALENO
Institute of Geography and Spatial Planning (IGOT), Universidade de Lisboa, Portugal

ABSTRACT

Following comparative historical analyses undergone over 20 years of research, both in developed and developing countries from Europe, Africa, Asia and Latin America, the question is: Can you afford to continue building unsustainable cities, under the current climate change challenges? Green corridors were implemented within the municipality of Lisbon, over the last decade, permitting residents and visitors to stroll, ride a bike, and grow their own food in several parks or community gardens approved by the Municipal Council and Assembly. The paper presents two case studies of community gardens established in the capital city of Portugal where food crops, spices and medicines are produced using organic waste from the gardeners' kitchens, making their households more resilient to climate change and less vulnerable to supply chains, thus reducing the carbon footprint for Lisbon residents. Hope is to contribute to urban sustainability science with good examples of environmental practices within cities and towns, towards a better future for us all.

Keywords: urban sustainability, Lisbon, urban agriculture.

1 INTRODUCTION

Community cultivation utilizes a combination of close spacing, intercropping and organic fertilization practices. According to Wade [1] this biodynamic intensive method of gardening begun in Europe in the 1800s, particularly in France. However, the Industrial Revolution in England led to the increase in urban food gardens, in order to address survival needs of incoming households [2], as rural migrants possessed agricultural skills and easily took care of vacant urban land, whenever and wherever they were allowed to. This is what happens in the 21st century Lisbon municipality, a global and cosmopolitan European metropolis, where the municipal legislation permits interested urbanites to grow food crops in the vicinity of their home, under a 80 to 146 euros annual fee, for plots of 50 to 120 m^2, thus providing access to land, compost tanks, and water, under the pre-condition that gardeners don't plant trees or raise poultry inside the community garden.

In fact, food gardening was always widely practiced in Lisbon and in its metropolitan region as stated in registers [3], usually in private front and backyards, in public schools and in museum parks, as it is intrinsically a Portuguese cultural approach to poverty, unemployment, underemployment and it is praised as educational to all generations [4]. However, Lisbon lacked the legislation and the municipal organizational background to make the activity possible and socially acceptable, until the 2000s, because farming activities are usually regarded as oxymoron in urban areas. From 2007 onwards, green corridors were implemented within the municipality, permitting residents and visitors to stroll, ride a bike, and grow their own food, spices and medicines in several food parks or community gardens approved by the Municipal Council and Assembly [5].

Around this time, Ruchelman was publishing the second edition of his book about American cities, entitled *Cities in the Third Wave* [6]. He argued that ideal compact cities could be efficient to keep, but they consumed large amounts of resources. Therefore,

WIT Transactions on The Built Environment, Vol 210, © 2022 WIT Press
www.witpress.com, ISSN 1743-3509 (on-line)
doi:10.2495/ARC220091

demands for clean air, open and green spaces, contact with nature and birds, were completely lost in agony with high levels of energy consumption, pollution driven transportation networks that generated lung diseases. Global cities were increasingly disconnected from the rural realm and urbanites had to travel long distances to look for calm and peaceful unpolluted environments. In the next decade, the housing crisis deepened socio-territorial inequalities in the United States but also in Europe, where Airbnb short-term rental market overheated the economy of cities such as Barcelona, Berlin, Amsterdam and Lisbon [7]. Despite the evictions of poor and elderly residents from Central Lisbon neighbourhoods, the municipal planning for greenways continued, as it was a strategy to uncompact the capital city of Portugal, as well as a movement intended to beautify boulevards and avenues, making the city a less polluted realm.

According to the Spanish geographer Hernández [8], both the serious social effects of fiscal consolidation, due to the sovereign debt crisis in Europe, and the appearance of extremist political parties, most of them green parties, gave way to the resurgence of climate change movements. These obliged the traditional conservative political parties to review their positions related to the design of cities, in order to mitigate the effects of a wave of extreme climate events. Other inorganic social movements forced this urban sustainability actions, such as the millennial environmental activist Greta Thunberg, who challenged world leaders, in the European Union and all the United Nations members, to change their policies related to the urban realm.

The objective of this work is to research ways to produce local food and decrease our carbon footprint. The idea is to look for sustainable management of soil, water and vegetable production within cities and metropolitan areas, under the label of urban agriculture. This is a long quest initiated in Latin America, continued in Africa and Asia, and improved in Europe, as part of "The Urban and Regional Change and Policies Research Group" (ZOE), from the University of Lisbon. Europe has wide environmental concerns, growing together with cultural and social cohesion policies, whose implementation marks the current European Commission agenda, the New Circular Economy Action Plan [9]. Under this plan, urban gardening contributes to reduce waste and ensure the implementation of the United Nations 2030 Sustainable Development Goals, particularly goal 11: Sustainable Cities and Communities [10].

The research question is: Can you afford to continue building unsustainable cities, under the current climate change challenges? Instead of continuing building compact cities that lack a sense of community, and neighbourhoods usually taken over by festival and culturally hungry tourists that have no other concerns but have a good time abroad, the hypothesis formulated is that we can envisage more environmentally sustainable cities if we accommodate rural activities in the urban realm, such as community gardens, integrated or not in greenways or green corridors (see Fig. 1). Another ambitious question is: are we building cities for its residents or as a business opportunity for future guests? Hope is to contribute for a better urban sustainability science.

2 MATERIAL AND METHODS

As the geologist Allègre stated [11], the study of the past is the key that permits us to predict the future. There is an ecological system in rural areas in contradiction with the ecology of the cities. A heat island covers metropolitan regions because the greenhouse gases capture temperatures that can be higher by 2–5° of the surrounding rural realm [12]. Natural balances can be gradually re-established, though, if instead of small greenery pockets, like parks and public gardens, planners introduce greenways and trails, nature conservation areas and

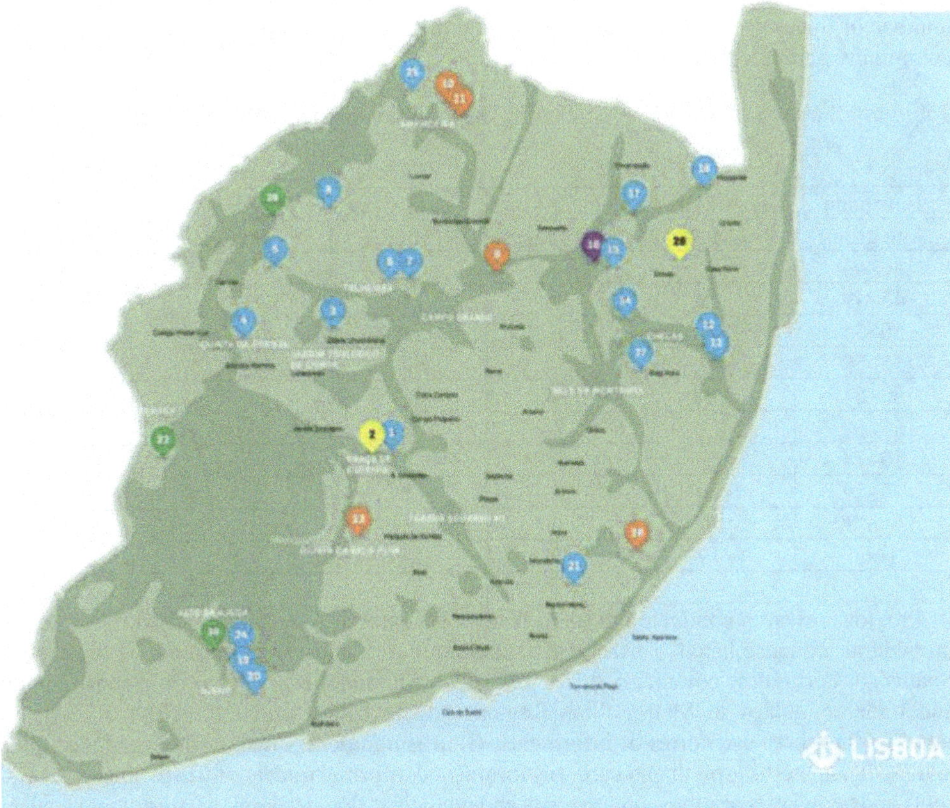

Figure 1: The municipality of Lisbon with the location of city parks and green corridors. The Bensaúde Horticulture Park, number 3 (in blue, as it has municipal management) and Padre Cruz community gardens, numbered 29 (in green, part of a greenway). The map reproduces current Lisbon municipality urban agriculture spots, 30 in total, and includes five parks and food gardens under non-municipal management (in orange) [15].

community gardens [13]. The possibility to lease public land for horticulture (cabbage, lettuce, tomatoes and carrots), for kitchen spices growth (rosemary and thyme), gives citizens access to affordable and fresh nutritious food that supplement market and supermarket purchases, whilst providing work to the urban poor and therapeutically driven activities for the elderly, enhancing environmental awareness and appreciation [14].

The current study was part of the Institute of Geography and Spatial Planning (IGOT) curricula, developed in urban geography classes of Honours degree students, and integrated in fieldwork techniques for geography learning of the Institute of Education (IE), and a Master's degree seminar also from the University of Lisbon. The students conducted surveys involving urban cultivators and gardeners, all around the metropolis, even though this contribution only focused on community gardens from the municipality of Lisbon, a total of thirty. Semi-structured interviews used the same questionnaire, intended to gather data about the age, sex (see Table 1), educational background, and profession of the urban gardeners, as well as their place of residence, type of transportation used to travel to the cultivated plot, number of hours spent weekly in farming activities, and the inputs to and outputs of the community garden cultivation.

Table 1: Surveys of Lisbon municipality community gardens. *(Source: Students 2019–2021.)*

Bensaúde	Horticulture	Park	Father Cruz	Neighborhood	Total
Age group	Male	Female	Male	Female	–
40–44	3	1	–	–	4
45–49	1	1	–	–	2
50–54	4	–	1	–	5
55–59	–	–	1	–	1
60–64	3	–	2	–	5
65–69	–	–	3	–	3
70–74	2	1	–	–	3
75–79	2	–	2	–	4
Over 80	2	–	1	–	3
Total	17	3	10	0	30

Previous research conducted in Latin America, Southern Asia and in Africa followed the procedural sequence listed: 1. Archival examination of manuscripts from the early years of Spanish or Portuguese colonization; 2. Fieldwork with random surveys to flora consumption and trade conducted in Metropolitan Regions and in Cities, under the form of in-depth interviews to three categories of informants: (i) urban gardeners responsible for food, spice and medicinal herbs growth and care; (ii) formal and informal traders of fruits, spices, tubers, roots and herbal preparations, pharmacists and medicinal flora traders; (iii) traditional healers and herb doctors that cure diseases and mild affections with potions of herbs or simples. Table 2 presents the number of testimonies conducted, mostly in tropical environments, and the proportion of urban agriculture gardeners and farmers interviews collected for two decades [16].

3 RESULTS

Urban agriculture (UA) is an activity that contributes to the production, processing and trade of food inside city limits, practiced using recycled organic wastes in compost tanks, an intensive ecological system that provides oxygen and biodiversity [18]. UA improves household's food security and quality within cities and metropolitan areas, giving income to the unemployed and underemployed urbanites, physical and mental health to ageing citizens, whilst improving family inclusion and integration in urban neighbourhoods, and in peri-urban towns [19]. The first survey focusing on urban gardening was conducted in the African city of Pointe-Noire (Congo), by the geographer Vennetier [20]. The 54,000-inhabitant city had about 3,000 households devoted to food production, ranging from cassava, corn, peanuts and potatoes to a diversity of horticulture crops. During the 1990s, another research conducted in six Kenyan cities concluded that 62% of the 1,576 urban households interviewed by the Mazingira Institute cultivated the food they consumed [21]. Most of the gardeners (56%) were women.

In 1998, the Portuguese Tropical Research Institute concluded a survey in Belém, Brazil, that accounted for 69.7% of female gardeners [22]. This is not the case with Lisbon, Portugal,

Table 2: Urban agriculture across the globe and across history [17].

Cities and Metropolis	Interviews (No.)	% Urban gardeners	Research mission years	Country
Belém, Pará	570	97.3	1998, 2005	Brazil
Presidente Prudente	280	100	1999	Brazil
Santarém, Pará	50	64.0	2006, 2007	Brazil
S. Luís, Maranhão	100	90.0	2010	Brazil
Recife, Pernambuco	50	6.0	2019, 2020	Brazil
Santiago Metropolis	132	84.1	2002, 2003, 2005	Chile
Iquique	75	64,0	2012	Chile
Metro Central Mexican	100	42,0	2004, 2006	Mexico
Lima Metropolis	34	0,0	2004, 2016	Peru
Piura	40	0,0	2016	Peru
Havana	50	94.0	2009	Cuba
San José	43	69.7	2010	Costa Rica
Rio Cuarto	100	80.0	2011	Argentina
Colonia del Sacramento	50	48.0	2011	Uruguay
Panama City	50	6.0	2017	Panama
Maputo Metro Area	20	100	2000	Mozambique
Accra	50	4.0	2017, 2018	Ghana
S. Tomé	50	20.0	2019	S. Tomé
Kochi	50	80.0	2013	India
Goa	50	52.0	2014	India
Malacca	20	20,0	2013	Malaysia
Manila	50	2.0	2015	Philippines
Total or average	2,014	51.0	–	–

where only 10% of the 30 community gardeners interviewed in 2019 and 2021, were female (Table 1). About 60% of the Lisbon community gardeners are over 60 years old, also in opposition with urban farmers interviewed in another Latin American city, Santiago, Chile, where 40.2% of the 111 La Pintana municipality residents questioned in 2002 and 2003 (Table 2) were in the same age group [23]. However, the sample of the tiny Colonia de Sacramento, Uruguay (15,934 inhabitants, in 2011) registered a proportion of 58% urban gardeners above 60 years old.

Again, a noticeable difference signalled between the first and the last sample conducted on the issue of urban agriculture and developed across time is the much smaller proportion of part-time gardeners: about 54.3% were registered in late 1990s in Brazil against 40% in 2019–2021 in Lisbon, Portugal. Across time, research conducted in three continents revealed that UA doesn't provide self-sufficiency in grains (wheat, corn), tubers (potatoes) or rhizomes (cassava) but feeds families with fresh vegetables and fruits, as fruit trees and horticulture are part of green belt municipal programs in countries such as Argentina, Brazil, Chile, Cuba and Mexico, in Latin America; Horticulture integrates basic school curricula in Manila, the Philippines, Asia [24]; Peri-urban horticulture and poultry was technically and financially sponsored in Maputo, Mozambique, in the year 2000 [4].

The 22 case studies numbered in Table 2 also gave evidence of the importance of aromatic and curative herb production in front and backyards, as well as in peri-urban farming plots, whose growth contributes to improve the health of the most vulnerable citizens, mostly using native flora already mentioned in early colonization year manuscripts [17]. The cities that registered a higher number of species applied externally or taken in teas and potions were: S. Luís, Brazil (109 different taxa); Río Cuarto, Argentina (125 species); and Kochi, India (136 plant species, either grown or traded in Ayurvedic pharmacies). The number of plants grown in Lisbon community gardens was quite small, as the 2019–2021 survey found only seven species used for medicinal purposes: Peppermint (7); Rosemary (four occurrences); Melissa (4); Lemongrass (six occurrences); Lavender (2); Sage (2); and *Lippia citriodora* (3).

3.1 Characterisation of Padre Cruz community gardens

The Carnide neighbourhood plots are 50–100 m^2 sized (5/20 m), close fenced and possess water for irrigation and horse manure, as well as compost tanks, for fertilization purposes (see Fig. 2). Only one community gardener was a part-time construction worker that travelled from another municipality (Amadora), on weekends, to water and tend his 100 m^2 plot, but the remaining nine were retired males, resident in two closely located social housing blocks, named Padre Cruz and Horta Nova. The gardeners travelled daily, walking or by public transportation, except for 40% of the 10 interviewed tenders that brought their car three times a week (Table 1). As to the educational background of Padre Cruz gardeners: 30% were illiterate; 40% had only basic schooling (4 years of studies); one had completed secondary school (12 years of studies); and one had completed an honours degree.

Figure 2: Detailed aerial photo of Padre Cruz community gardens, with a clear image of the plots or food gardens [25].

The survey was conducted by an IGOT Geography student resident in Carnide neighbourhood (see Fig. 1), during his 2019 urban geography schoolyear. Figs 2, 3 and 4 were part of his essay. Padre Cruz community gardens main productions were corn (20%)

and potatoes (30%), whilst the majority produced bush beans (90% of the plots), cabbage (30%), lettuce and onions (20% each), hot and sweet pepper (10%). Rosemary was also gardened, used as spice and as medicine. These community gardens are part of a green corridor located in the north-west of Lisbon municipality, connected to Monsanto City Park.

Figure 3: Horticulture community garden plot [25].

Figure 4: Beans tended in Padre Cruz gardens [25].

3.2 Characterisation of Bensaúde horticulture park community gardens

In contrast with Padre Cruz gardens, Bensaúde Park, located in the neighbourhood of Benfica, (south to Carnide area), constitutes the so-called doctors' food gardens, as about 60% of the tenants completed an honours degree, their professions varying from engineer, architect, lawyer, agronomy professor and accountant. Nowadays, about 40% are retired, and one gardener interviewed was unemployed, with food production having a vital role in his food security. It is interesting to emphasize that two municipal technicians, resident in the neighbourhood of Benfica, also tended their plot, as a part-time job and intended to supplement their household's diet. Bensaúde Park plots have an area that goes from 80, 100 to 120 m^2 and the rent paid is of 100, 125 to 146 euros per annum, respectively.

Bensaúde Park covers 3.5 hectares of Greenland, formerly farmed and part of the Lisbon greenbelt. The community gardens total 21, but one is empty. They are fenced, and have small shacks used to keep the utensils used, but not farming machines, as they aren't allowed in these municipal food plots. The gardens have compost tanks and water, small nurseries and some remaining trees, like almonds, shading some food gardened areas. The interviews were conducted by Master's degree students, during the Fieldwork in Geography Seminar, in the schoolyear of 2021. Table 3 shows the diversity of horticulture crops tended in this second example. Besides these, Bensaúde gardeners also grow potatoes and sweet potatoes, a diversity of herbs and spices (coriander, oregano, parsley, garlic, thyme, laurel, cumin, and savoury), medicines (peppermint, rosemary, lavender, sage, and mint) and fruits (passion fruits, strawberries, raspberries, and melons).

4 DISCUSSION

A total of 45 different plant species was recorded in both food gardens, in 2019 and 2021. Change in dietary habits all over Europe, where veganism is growing as environmental awareness increases, have brought back vegetable gardens from peri-urban to urban areas [26]. The sustainability of urban ecosystems, and the western romanticized picture of

Table 3: Horticulture crops from Bensaúde Park community gardens. *(Source: Master's students essays, 2021.)*

Common name	Botanic name	Use	Occurrences
Kale	*Brassica oleracea* var. *acephala*	Green leaf	1
Pumpkin	*Cucurbita pepo*	Cucurbit	2
Lettuce	*Lactuca sativa*	Green leaf	14
Leek	*Allium ampeloprasum* var. *porrum*	Bulb	5
Beetroot	*Beta vulgaris*	Green leaf, root	1
Onion	*Allium cepa*	Bulb	14
Eggplant	*Solanum melongena*	Fruit vegetable	1
String bean	*Phaseolus vulgaris*	Legume	1
Cucumber	*Cucumis sativus*	Fruit vegetable	1
Pepper	*Capsicum annum*	Fruit vegetable	1
Broccoli	*Brassica oleracea*	Brassica	2
Carrot	*Daucus carota*	Root	3
Cabbage	*Brassica oleracea* var *capitata*	Brassica	16
Pea	*Pisum sativum*	Legume	6
Spinach	*Spinacia oleracea*	Green leaf	1
Turnip	*Brassica rapa*	Green leaf, root	4
Tomatoes	*Lycopersicum esculentum*	Fruit vegetable	4
Fava bean	*Vicia faba*	Legume	11
Okra	*Abelmoschus esculentus*	Fruit	1
Celery	*Appium graveolens*	Whole plant	2
White lupine	*Lupinus albus*	Grain pulse	1

individuals feeding the city, explain the diversity in horticulture production in Bensaúde Park community gardens (see Table 3), tended by highly educated residents of Benfica neighbourhood, but not entirely. In fact, a total of 21 different vegetable species are found in 80 to 120 m^2 plots, showing how local food supply can be enriched in quality and quantity, inside a European capital city.

Turner et al. [27] wrote that community gardens are quite diverse elsewhere around the world, including: "city farms, therapeutic gardens, school kitchen gardens, or a more individual allotment system" [27]. These authors have analysed the European History of UA, and they stated that in times of crisis citizens always felt encouraged to play an active role in their own food production, as registered during the two World Wars, the 1970s oil crisis, and the collapse of the Soviet Union. So, it is quite unsurprising that, during the current pandemic CORONA virus outbreak, the Bensaúde Horticulture Park allotments registered intensive cultivation, even amid more wealthy urbanites.

The Padre Cruz community gardens, researched just before this pandemic outbreak (2019), are food-based social-allotment gardens, associated to less-wealthy Lisbon residents, long concerned about their household food-security. This case study constitutes a good example of economic and environmental inclusion, as it is in a green corridor also sought after by neighbouring illegal settlement residents, like Cova da Moura builders, as found during the survey. Morgan [28] argues that food security has long became a national security issue, hurried by catastrophic climate change effects that are giving way to mass migrations and escalating land conflicts. Food planning can no longer afford to leave urban areas out of the equation of how to get essentials for human life existence on planet earth. Empowering urban residents to produce their own food, either by creating individual allotment gardens or by allowing them to grow food in terraces and walls is a challenge for national and municipal governments. "Locally produced food is the most ecologically sustainable because it has lower food miles", therefore reducing our "carbon footprint" [28].

5 CONCLUSIONS

Following the comparative historical analyses undergone over twenty years of research, the hypothesis formulated for the current study project was that we can envisage more environmentally sustainable cities if we accommodate rural activities in the urban realm, such as community gardens, integrated or not in greenways, instead of continuing to build compact cities that lack a sense of community. Lisbon community gardens were found to have a good potential to promote social inclusion, particularly Padre Cruz allotments, tended by social housing residents from two nearby built areas (Padre Cruz and Horta Nova) and sought after neighbouring Amadora municipality residents, from illegal migrant areas such as Cova da Moura. Community-building potential was also found in both community gardens, the less-wealthy Carnide greenway and the wealthier Benfica neighbouring horticulture park.

The integration of allotment gardens in green corridors was a beneficial innovation from Lisbon municipal government, as in the year 2020 the city received the prize for European Green Capital, attributed by the European Commission. With a population of over 500,000 inhabitants, the municipal government was recognized to have a cohesive city-wide vision for sustainable mobility and to have achieved that through measures to restrict car use and prioritize walking, cycling and public transport. Lisbon also connected its green areas with the Alcântara valley green corridor and linked up Monsanto Park and the Tagus Riverine paths, with cycling and walkways, giving its citizens greater access to green spaces. The tale of two community gardens, one of them integrated in Monsanto Park green corridor is a sound contribution to improve micro-climatic conditions and reduce the heat island effects, as well as of improving waste recycling management and reducing food miles, by localizing food production and giving access to soil prepared for gardening purposes.

REFERENCES

[1] Wade, I., *City Food: Crop Selection in Third World Cities*, Urban Resources System: San Francisco, 1986.

[2] Mateus, M.L.P.F.R., Campos de Coimbra. Do Rural ao Urbano. Unpublished PhD thesis, Universidade de Coimbra: Coimbra, 2009.

[3] Cavaco, C., A pequena horticultura doméstica de não agricultores das periferias urbanas: O caso de Lisboa. *Economia,* **VIII**(2), pp. 395–423, 1984.

[4] Madaleno, I.M., Urban agriculture, an environmentally sustainable land use: Case studies from Europe, Latin America and Africa. *WIT Transactions on Ecology and the Environment*, vol. 129, WIT Press: Southampton and Boston, pp. 961–969, 2002. DOI: 10.2495/URS020921.

[5] Castro-Henriques, J., Urban agriculture and resilience in Lisbon: The role of the municipal government. *Urban Agriculture Magazine*, **22**, pp. 49–50, 2009.

[6] Ruchelman, L.I., *Cities in the Third Wave: The Technological Transformation of Urban America*, Rowman and Littlefield: New York, 2007.

[7] Mendes. L., Transnational gentrification and the housing market during pandemic times, Lisbon style. *Urban Geography online*, 2021. DOI: 10.1080/02723638.2021.1949832.

[8] Hernández, J.L.S., Los mecanismos organizativos de las prácticas económicas alternativas en las ciudades españolas. *Documents d'Anàlisi Geogràfica,* **67**(3), pp. 417–440, 2021. DOI: 10.5565/rev/dag.658.

[9] European Commission, Circular Economy Action Plan: For a cleaner and more competitive Europe. Interreg Europe: Brussels, 2020. https://ec.europa.eu/environment/pdf/circulareconomy/new_circular_economy_action_plan.pdf. Accessed on: 5 Jul. 2021.

[10] United Nations, The SDGS in action. United Nations Development Programme: New York, 2021. https://www.undp.org/sustainable-development-goals. Accessed on: 5 Jul. 2021.

[11] Allègre, C., *Ecologia das Cidades, Ecologia dos Campos*, Instituto Piaget: Lisbon, 1996.

[12] Spiegelhalter, T. & Arch, R.A., Biomimicry and circular metabolism for the cities of the future. *WIT Transactions on Ecology and the Environment*, vol. 129, WIT Press: Southampton and Boston, pp. 215–226, 2010. DOI: 10.2495/SC100191.

[13] Wolch, J.R., Byrne, J. & Newell, J.P., Urban green space, public health, and environmental justice: The challenge of making cities "just green enough". *Landscape and Urban Planning*, **125**, pp. 234–244, 2014.

[14] Lautenschlager, L. & Smith, C., Beliefs, knowledge, and values held by inner-city youth about gardening. nutrition, and cooking. *Agriculture and Human Values*, **24**, pp. 245–258, 2007. DOI: 10.1007/s10460-006-9051-z.

[15] CML, Mapas do Município de Lisboa. Câmara Municipal: Lisboa, 2021. https://www.google.com/search?source=univ&tbm=isch&q=Mapas+do+munic%C3%ADpio+de+Lisboa. Accessed on: 2 Nov. 2021.

[16] Madaleno, I.M., Spices and medicines cultivated and traded in Panama-City. *La Geografía de las Redes Económicas y la Geografía Económica en Rede,* eds M.P. Alonso Logroño, T.S. Marques & H. Santos, Universidade do Porto and Asociación de Geógrafos Españoles: Porto, pp. 249–257, 2020. DOI: 10.21747/9789898969460/geoa24.

[17] Madaleno, I.M., Etno-Geografía de Flora Útil de Latinoamérica. Un análisis histórico comparativo de los siglos XVI y XXI. Hacia una perspectiva integrada. Aportaciones de las Geografías Económica, Regional y de los Servicios para la cohesión y la competitividad territorial, eds C. Gago, J. Córdoba, M.P. Alonso, R.M. Jordá & J. Ventura, pp. 63–67, 2021.

[18] UNDP, *Urban Agriculture: Food, Jobs and Sustainable Cities,* United Nations Development Programme: New York, 1996.

[19] Bakker, N., Dubbeling, M., Gundel, S., Sabel-Koschella, U. & Zeeuw, H., *Growing Cities, Growing Food,* German Foundation for International Development: Feldafing, 2000.

[20] Vennetier, P., La vie agricole urbaine à Pointe-Noire (Congo). *Cahiers d'Outre Mer*, **14**(53), pp. 60–84, 1961.

[21] Sawio, C.J., Urban agriculture research in East and Central Africa: Record, capacities and opportunities. *Cities Feeding People*, IDRC: Ottawa, pp. 1–26, 1993.

[22] Madaleno, I.M., Urban agriculture in Belém, Brazil. *Cities*, **17**(1), pp. 73–77, 2000.

[23] Madaleno, I.M. & Gurovich, A., "Urban versus rural" no longer matches reality: Endurance of an early public agro-residential development in peri-urban Santiago, Chile. *Cities*, **21**(6), pp. 513–526, 2004.

[24] Madaleno, I.M., Food and medicinal plants consumed in Manila, The Philippines. *WIT Transactions on the Built Environment*, vol. 170, WIT Press: Southampton and Boston, pp. 203–210, 2017. DOI: 10.2495/CC170201.

[25] Abreu, A., Hortas Urbanas Carnide. Unpublished Honours degree essay, Universidade de Lisboa: Lisbon, 2019.

[26] Herchenbach, M., Food and the city. *Journal of Landscape Architecture*, **7**(2), p. 89, 2012.

[27] Turner, B., Henryks, J. & Pearson, D., Community gardens: sustainability, health and inclusion in the city. *Local Environment: The International Journal of Justice and Sustainability*, **16**(6), p. 489, 2011. DOI: 10.1080/13549839.2011.595901.

[28] Morgan, K., Feeding the city: The challenge of urban planning. *International Planning Studies*, **14**(4), p. 345, 2009. DOI: 10.1080/13563471003642852.

CONTRIBUTION OF THE BIOPHILIC DESIGN APPROACH TO THE UN SUSTAINABLE DEVELOPMENT GOALS

IMAN ABDELSHAHID IBRAHIM & BALSAM MUNADHIL AL-CHADERCHI
[1]College of Fine Arts and Design, University Of Sharjah, United Arab Emirates
[2]Architecture Department, University Of Sharjah, United Arab Emirates

ABSTRACT

In the last decade, biophilic design under the concept of creating strong connections between nature and manmade environments have spread significantly. Because of the rapidly increasing environmental crises and climate change phenomena globally. Consequently, biophilic design has become associated with sustainable applications, the question this research discusses is if this approach contributes to achieve the aspects of sustainability, under the umbrella of the UN Sustainable Development Goals (SDGs). This study will investigate biophilic approach contribution to sustainable design through defining aspects of nature that most influence human satisfaction of built environment. Aligned with the European Commission (2015) series actions titled "nature-based solutions" that are inspired by nature systems. During the COVID-19 lockdown, humans had minimal interaction with nature, and access to green public spaces, at the same time, an increasing interest to integrate nature in design, reducing humans impact on nature were recognized in the contemporary built environment. The study adopts William Browning's (2014) "14 Patterns of Biophilic Design" framework, that lays out a series of tools experiencing science and nature contribution for promising opportunities creating sustainable built environment design applications. In addition to identifying and analysing the current practices for biophilic design in the built environment and its' impacts on nature and human wellbeing. The results conduct design strategies, and guidelines that integrate biophilic approach as a base, within the 14 patterns framework for better practices creating innovative and responsible built environment.
Keywords: integration, biophilic design, sustainable built environment, nature-based solutions.

1 INTRODUCTION

The link of "nature" and human settlements has a deep history. The fabulous Hanging Garden of Babylon is thought to have been a brilliant construction in classical antiquity that was adjacent to the water source and filled with a rich diversity of trees, shrubs, and grapevines in terraced gardens. In modern architecture, architects discover living with nature through a wider range of approaches. For example, Leberecht Migge planned the installation of edible gardens. Apartments with private gardens were included in Le Corbusier's conceptual project, Immeubles villas [1]. Biophilic design (BD) can reduce stress, enhance creativity and clarity of thought, improve our well-being, and expedite healing; as the world population continues to urbanize, these qualities are ever more important. Theorists, research scientists, and design practitioners have been working for decades to define aspects of nature that most affect our satisfaction with the built environment. Several publications studied different approaches to biophilic design. The air approach by Browning et al. [2], Gou et al. [3] and Kellert [4] suggests to increase natural ventilation using operable windows, vents, narrower structures, etc. and simulate natural air and ventilation through operable windows, vents, airshafts, porches, clerestories, HVAC systems, etc. The water approach by Browning et al. [2] and Kellert [4] suggests building waterscapes such as fountains, constructed wetlands, ponds, water walls, rainwater spouts, aquaria, etc. and access to natural water features such as waterfalls, rivers, streams, oceans, etc. The light approach by Browning et al. [2] and Kellert [4] suggest bringing in natural light via glass walls, clerestories, skylights, atria, reflective colours/materials, etc. and mimic the spectral and ambient qualities of natural light,

WIT Transactions on The Built Environment, Vol 210, © 2022 WIT Press
www.witpress.com, ISSN 1743-3509 (on-line)
doi:10.2495/ARC220101

such as by arranging multiple low-glare electric light sources, ambient diffused lighting on walls/ ceiling, and daylight preserving window treatments. The plants approach by Chang and Chen [5] suggest bringing vegetation indoors by potting plants and indoor green walls and incorporate plants into buildings by using green roofs, green walls and facades, large atria with park-like settings, green pockets, etc. Another approach by Schweitzer et al. [7] suggests to create spaces to accommodate animals, such as ponds, aquariums, etc. and wild animal-friendly living areas to attract animals like nest boxes, gardens, green roofs/walls, etc. Additionally, Browning et al. [2] and Fei et al. [6] mention an approach involving landscapes in the sites such as constructed wetlands, grasslands, prairies, forests, and other habitats, interior landscapes in atria, courtyards, entry areas, hallways, etc. Also providing window views of natural landscapes like forests, seascapes, and water motifs. The study adopts William Browning's (2014) "14 Patterns of Biophilic Design" framework shown in Fig. 2, that lays out a series of tools experiencing science and nature contribution for promising opportunities creating sustainable built environment design applications. These patterns are comprehensive and include all the elements of nature that are encouraged by the many publications that has been reviewed.

Figure 1: The Sustainable Development Goals (SDGs) or global goals shown are a collection of 17 interlinked global goals designed to be a "blueprint to achieve a better and more sustainable future for all" [9].

According to European Commission (2015), nature-based solutions aim to help societies address a variety of environmental, social, and economic challenges in sustainable ways. They are actions inspired by, supported by or copied from nature. Nature-based solutions use the features and complex system processes of nature, such as its ability to store carbon and regulate water flows, in order to achieve desired outcomes, such as reduced disaster risk and an environment that improves human well-being and socially inclusive green growth [8].

2 OBJECTIVES AND METHODOLOGY

This paper is a step towards implementing sustainable biophilic design, and it aims to identify the sustainability strengths and weaknesses of BD patterns. It presents a review of projects, with the goal of identifying its biophilic features and their sustainable impacts. This study

Nature in the space

- **[P1] Visual connection with nature:** a view to elements of nature.
- **[P2] Non-visual connection with nature:** auditoryreference to nature.
- **[P3] non-rhythmic sensory stimuli:** stochastic connections with nature
- **[P4] Thermal and airflow variability:** mimic natural environments.
- **[P5] Presence of water:** the seeing, hearing or touching of water.
- **[P6] Dynamic and diffuse light:** varying light to create nature conditions
- **[P7] Connection with natural systems:** awareness of seasonal changes

Natural analogues

- **[P8] Biomorphic forms and patterns:** texture that persist in nature.
- **[P9] Material connection with nature:** reflect the ecology or geology
- **[P10] Complexity and order:** spatial hierarchy similar to nature

Nature of the space

- **[P11] Prospect:** an unimpeded view over a distance for surveillance
- **[P12] Refuge:** a place for withdrawal, from environmental conditions
- **[P13] Mystery:** the promise of more information
- **[P14] Risk/Peril:** an identifiable threat coupled with a reliable safeguard

Figure 2: The biophilic design patterns within three categories based on nature relationship [10].

also reveals the potential sustainable benefits of adopting BD and the existing challenges that must be overcome, ultimately paving the way for the enhancement of human nature connection within sustainable built environments. In addition to conduct design strategies, and guidelines that integrate biophilic approach as a base, within the 14 patterns framework for better practices creating innovative and responsible built environment.

2.1 Methodology

The research methodology structured to analyse selected case studies represents current practices for BD projects in relation to SDGs achievements, with certain criteria included identification, screening, and eligibility, as shown in Fig. 3.

2.2 Case studies selection criteria

Based on the study limitation to analyse the BD case studies, there are three steps for considering the selected current practices for BD, started by identification, followed by screening and ending with eligibility as follows:

2.2.1 Step 1: Identification

The keywords used for the search were "biophilic", "biophilic design", "biophilic architecture", "biophilic patterns". The design concept was the first exclusion criterion. The fields included were architecture, interior design and furniture. Fields such as products, manufacturing, biology, and land conservative were not relevant subject areas for this study and were therefore excluded.

2.2.2 Step 2: Screening

In the second phase, all of the cases were reviewed to refine the selection list, and the second exclusion criteria was used to further narrow the selection of cases for eligibility is claiming to be "sustainable" or "green". This is a crucial exclusion point because each project would be judged to see if the biophilic patterns or features are sustainable or just claiming to be green two biophilic design.

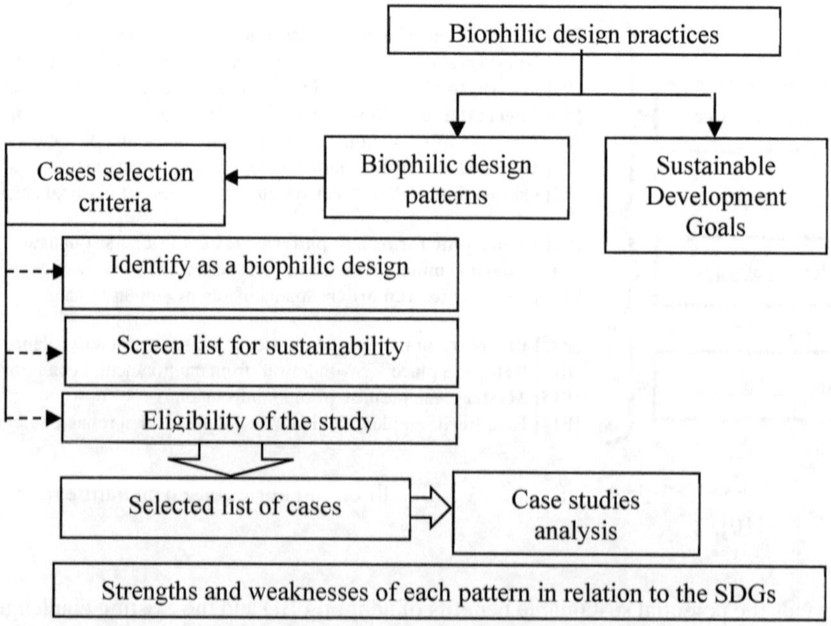

Figure 3: Research methodology chart.

2.2.3 Step 3: Eligibility

The goal of the third step was to filter the remaining cases to identify the projects with the highest potential for this study. All the documents were filtered according to two points. The two filtering criteria used to assess their potential applicability to the scope of the study were: focus on the Realistic projects not educational or competition and focus on 500 m^2 or more. By the end of this phase, each project taken under consideration is or has by biophilic patterns, claiming to be sustainable, realistic, completed and is 500 m^2 or more.

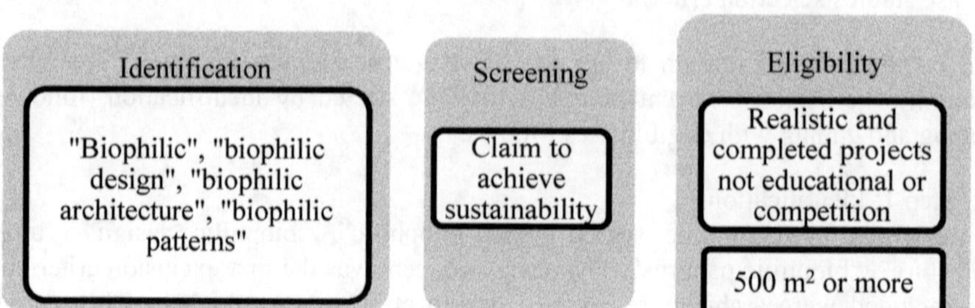

Figure 4: Selection criteria for BD case studies.

3 SUSTAINABLE DEVELOPMENT GOALS EVALUATION GUIDELINES

Investigating the eligibility of some biophilic design to have sustainable impact and contribute two sustainable aspects requires to set some guidelines and answer question "does the built environment meets any of these aspects?" in this case UN Sustainable Development

Goals SDGs are the guidelines for contributing to achieve the aspects of sustainability. Under the umbrella of sustainability, the 17 SDG cover a broad spectrum of causes like social issues so we will focus in this study on the environmental and design issues thus we will only choose eight goals shown in Fig. 5 out of the 17 to realize if these case studies will achieve any of them.

Figure 5: Study limitation to selected SDGs realized within the BD case studies selected.

4 BIOPHILIC DESIGN SELECTED CASE STUDIES ANALYSIS

4.1 Greenacre Park

Designed by Sasaki, Dawson, DeMay and Associates, in New York, USA. Within an area of 6,000 ft^2 (557 m^2), and completed in 1971.

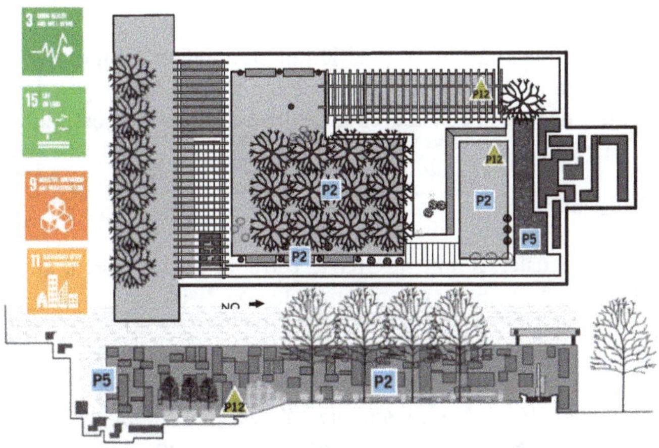

Figure 6: Greenacre Park layout and section drawing showing the BD patterns application and SDGs application [10].

4.1.1 Greenacre Park's strengths

The case presents with several biophilic features that achieve sustainable UN Sustainable Developments Goals. Abundant trees, shrubs, smell and texture of stone natural material contribute to the productivity of individuals and their mental health thus creating better community and more efficient environment to live and work in. The high-density spots of the plants help mimic the actual experience of being in nature in addition to the waterbodies contribution to visualizing and hearing the experience of being inside a forest or simply a greenery outdoor environment (**[P1] [P2] [P9] → SDG 3**). Rustling ivy on the west wall and tree canopy above, and birds contribute to sustainability of the surrounding living creatures and their flourish will help to not only maintain the creature but also to maintain the environment that we all coexist in it. The incorporation of this feature is highly advanced approach to help humans embrace the nature in the environment rather than eliminating nature to create a solitary environment (**[P3] [P7] → SDG 15**). Cool air wafts up from the water features, shading provided by trees, movable furniture ensures preferred conditions that reduce reliance on active systems. This feature is considered a passive approach to create thermal comfort and reduce use energy consumption. In addition, the comfortable environment with adequate ventilation and easy access layout promotes health, wellbeing, and stress-free zone (**[P4] → SDG 11**). Light filters in through tree canopy; lower, upper, and middle levels experience different amounts of infiltration contribute to an aesthetic indoor experience. Such application motivates the human mind and stimulate the sensors specially in artistic spaces allowing innovation and the attraction to observe the changes in light and the depth of the details of the natural materials and design (**[P6] [P12] [P13] → SDG 9**).

4.1.2 Greenacre Park's weaknesses

Although water bodies are encouraged in the space to provide cooling and connection to nature but a 25 ft tall waterfall cascades down the north wall is extremely consuming to the environment. The water source has not been established that it comes from renewable source, or it has been recycled such as using greywater and the flow of the waterfall is not stated to be energized by a renewable source which make it a weak and opposite to sustainability and conservation of resources. In addition, having a stone sculpture can be considered over the top, this addition to the space can waste of natural materials that are precious to our environment and must be used sustainably **[P5]**. Elevating the park above the street level must be for technical purpose and for a specific function; either for the structural integrity or for the safety of conditions such as rain but creating this elevation just for the purpose of aesthetics or biophilic design is considered an extra construction cost, time, materials, and energy **[P11]**.

4.2 Glumac's Shanghai office

Designed by Glumac, Gensler, Shimizu, GIGA, Terrapin Bright Green, in Shanghai, China. Within an area of 10,000 ft^2 (929 m^2), and completed on 2014.

4.2.1 Glumac's Shanghai office's strengths

Indoor plantings, garden views and partially obscured view of plants at the end of the long hallway. All these patterns help physical and mental health of a human being because existing in nature is calming thus it reduces stress. Integrating nature in and to the built environment can minimize anxiety, overthinking and pressure which effects both happiness and

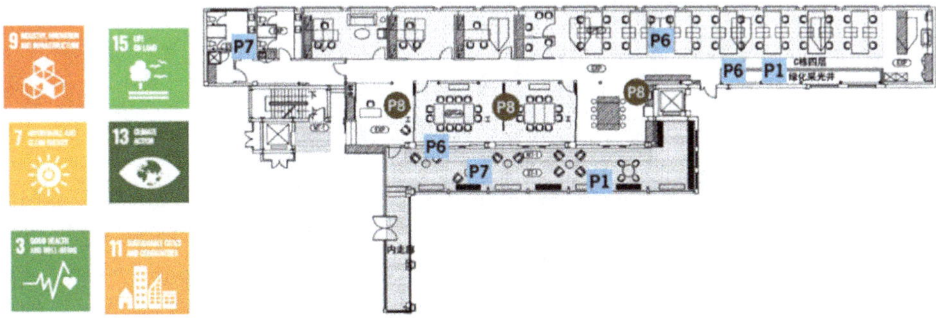

Figure 7: Glumac's Shanghai office plan drawing showing the BD patterns application and SDGs applied [10].

productivity of the individual (**[P1] [P13] → SDG 3)**. The pleasant odour of indoor plants, the sounds of the birds that inhabit the garden courtyard encourage the flourish of the natural environment that surrounds the built environment. This connects the outdoor to the indoor and create movement energy in the space and minimize the impact of the built environment on the surrounding (**[P2] → SDG 15)**. Fully operational windows throughout the space, and doors to the patio which can be propped open provides fresh air and daylight creating natural ventilation and natural illumination in the space thus reducing the need for energy consuming active systems such as HVAC and artificial lights. Open office plan and an elevated long-distance view from the patio added resilience to the space and to the building making it more comfortable and durable increasing the space life span and making it adaptable to multiple functions (**[P4] [P11] → SDG 11)**. A light well and large windows flood the space with daylight; adjustable transparency of the glass wall provide variability resilience of the functions and the buildings spaces because well lighted spaces can be used for anything and does not rely on lighting equipment's to achieve the desired illumination for different tasks that the space is intended. Open office plan and an elevated long-distance view from the patio contribute to the possibility of retrofitting the building and upgrading its potential overtime (**[P6] [P11] → SDG 9**). Reclaimed wood reception desk and bistro table, unpainted strawboard columns and cabinets and reclaimed Chinese gray bricks reduces greenhouse gases by reusing products without processing them which is a more responsible and sustainable method than recycling which requires processing thus contributes to climate change (**[P9]→ SDG 13**).

4.2.2 Glumac's Shanghai office's weaknesses

A light well and large windows flood the space with daylight; adjustable transparency of the glass wall and individual task lighting provide variability and control. This approach might make a lot of thermal bridges which compromises the thermal comfort it needs careful considerations and calculations. The cost is also a place of question since and advanced technology like this might be too expensive if it is not made from recycled materials **[P6]**.

4.3 Kickstarter Offices

Designed by Team Ole Sondresen Architect, in New York, USA Within an area of 29,000 ft^2, and completed in 2014.

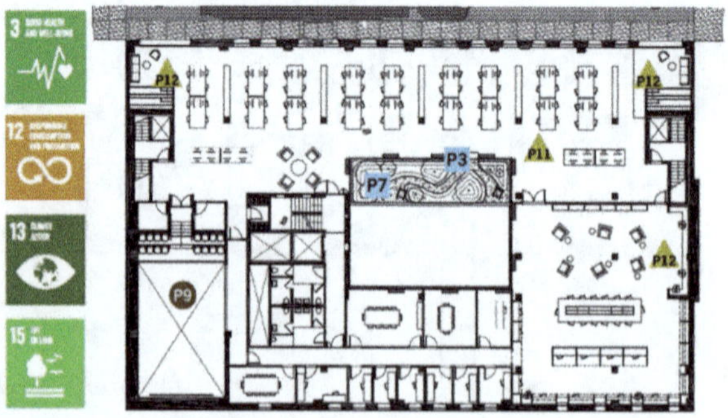

Figure 8: Kickstarter office plan drawing showing the BD patterns application and SDGs applied [10].

4.3.1 Kickstarter Offices' strengths

The views to the outdoors from all three floors and every work desk provides a sense of connection to the surrounding environment. Especially if the surrounding environment is green and have features such as moving grasses, edible berries, edible garden plots. Such an environment will assist in mental health of the workers (**[P1] [P2] → SDG 3**). Moving grasses, edible berries, edible garden plots represent a small scaled urban farming. The products of these edible garden slots are very important, and it provide a deeper link to the nature through taste sense. This quality is considered sustainable consumption because it does not depend on mass manufacturing and encourages the increase of functional planting (**[P2] → SDG**). The use of native landscaping sets the stage for non-rhythmic sensory stimuli, with occupants on each floor with direct view of native landscaping. In addition, it encourages the growth of local plants and making sure that the ecosystem remains for other creatures to be sustained for longer periods. The central courtyard and green roof's native landscaping rainwater capture and retention in the courtyard saves resources such as water and provides natural irrigation. The use of green roof is known to prevent heat island effect and creates a protection for the interior-built environment from outdoor conditions (**[P3] [P5] [P7] → SDG 15**). The operable windows in the office, natural light from the courtyard, sunroom, and exterior refurbished windows provides natural ventilation and natural illumination to the space thus reducing use of active system and the consumption of energy. Minimizing the use of active systems and relying on renewable energy such as wind and sun minimize the greenhouse gases and contribute to preventing climate change. Interior doors, desks and seating are made of salvaged or local materials, which extends the life cycle of used materials and saves energy of production, manufacturing and transporting new materials (**[P4] [P6] [P9] → SDG 13**).

4.3.2 Kickstarter Offices' weaknesses

The operable windows and open courtyard are appreciated but can cause indoor heat fluctuation due to thermal bridges. Such element is wanted but several factors must be considered like cost, recycled materials, amount of privacy and thermal comfort because too much use of light thermal mass such as glass leads to a weak envelope thus uncomfortable indoor condition. The point/aim of open spaces and windows is to allow naturally functional spaces but these spaces still must be controlled from the indoor **[P4] [P7]**.

4.4 Östra Hospital

A hospital in Göteborg, Sweden, designed by Team White Architects. Within an area of 193,750 ft^2 (18,000 m^2), and completed in 2006, the hospital considered one of the BD prototypes.

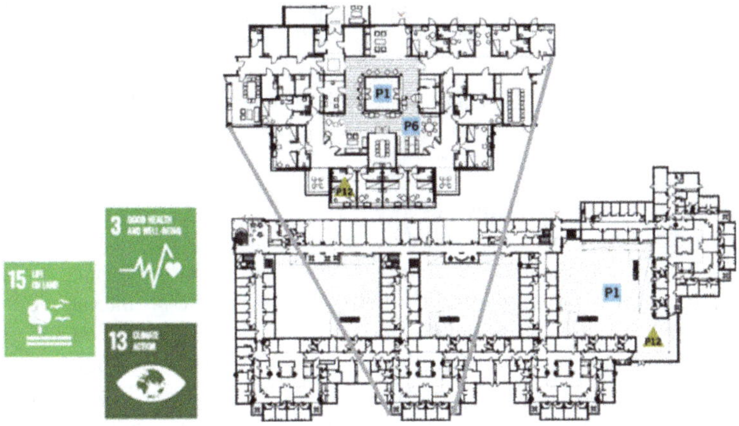

Figure 9: Östra Hospital plan drawing showing the BD patterns application and SDGs applied [10].

4.4.1 Östra Hospital's strengths

Patient rooms have views to the central gardens; light courts are planted with low vegetation providing a sense of connection to the surrounding environment and natural filtration/renewal of air indoors. Especially if the surrounding environment is green and have feature such as courtyards with edible plants. Such an environment will assist in mental health of the occupants (**[P1] [P2] → SDG 3**). Operable windows allow for responsible use of resources because it utilizes renewable energy such as sun and wind reducing the need to use artificial lighting and HVAC (**[P2] → SDG 12**). operable sunshades and windows light courts introduces natural light into the indoors does thus would you think the need for artificial lighting that consume energy (**[P4] [P6] → SDG 13**). Garden courtyards encourage biodiversity and create home for surrounding creatures that may lose the part of their palms to the built environment. In addition, it connects the natural environment to the built space (**[P3] [P7] → SDG 15**).

4.4.2 Östra Hospital weaknesses

Use of natural materials gives a nice addition to the interior but using non-recycled or non-reclaimed materials such as natural stone and wood increase the consumption of trees and invasion of forests and compromise the ecosystem thus compromising the creatures and the different natural components of the surrounding **[P9]**.

4.5 Parkroyal on Pickering Hotel

Designed by Team WOHA, in Singapore City, Singapore. Within an area of 12,121 ft^2 (1,126 m^2), and completed in 2013.

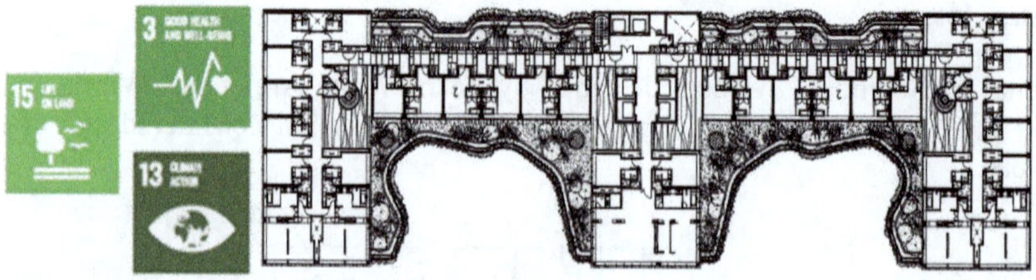

Figure 10: Parkroyal on Pickering plan drawing showing the BD patterns application and SDGs applied [10].

4.5.1 Parkroyal on Pickering's strengths

Views of Hong Lim Park or the sky-gardens are accessible from nearly every guest room; abundant indoor plantings and water features. Odours from indoor plants; sounds of birds and insects attracted by native plants. All these qualities create an interactive environment for the patients to break the norms and the boredom of staying in a hospital and gives a sense of calming and relaxation promoting health and healing (**[P1] [P2] → SDG 3**). Water features both inside and outside, light reflected off the water, natural light indoors and translucent curtains in guestrooms provides thermal comfort and sufficient lighting inside the interior space. Relying on such renewable resources to provide thermal comfort and to provide illumination reduce or eliminate active systems use thus minimizing the amount of greenhouse gases that cause climate change (**[P4] [P6] → SDG 13**). Planted sky-gardens attracts birds and other creatures which promotes biodiversity and sustainability of the natural life around the built environment. Also, it protects certain species and types of animals from migrating or dying by creating a sanctuary inside this space (**[P3] [P7] → SDG 15**).

4.5.2 Parkroyal on Pickering's weaknesses

Abundant water features incorporated into sky-gardens, lobby, passageways, infinity pool must be used in a responsible matter by recycling grey water, otherwise it considered a waste of water, and energy **[P5]**. Unfinished stone-walls, polished granite walls, and unpainted wood surfaces should be recycled reclaimed or reused because such precious components of nature are necessary to sustain the natural environment rather than be integrated into the built environment just for an aesthetic reason **[P9]**.

5 CONCLUSION

This paper identified and analyzed several projects that present many of the 14 biophilic patterns in terms of connection and relation to the nature-based solutions, that most influence human satisfaction of built environment. Where the research study investigated the biophilic approach contribution to achieve the aspects of sustainability, under the umbrella of the UN sustainable Development Goals SDG's. The patterns represented in each case have similarities to theoretical identification for biophilic approach, but the application showed clearly, how far they achieve some of the UN SDG's. It showed as well, how far application of some BD patterns could affect some sustainability aspects negatively, where decision here should compromise strengths against weaknesses. For instance, patterns use natural resources, as pattern number five "presence of water" is important as a wellbeing factor but can be a waste of precious resources like with huge energy operated, waterfall that also does not use recycled water for more efficiency. In such case, it puts a strain on natural resources in contradiction with SDG number 12 "responsible consumption and production". In such cases research recommend more sustainable tools for application and using clean resources like PV operated fountain that relies on grey water. The author believes that most biophilic patterns serve a sustainable purpose achieving many SDGs, increasing human satisfaction of built environment, connecting them more with nature. For advancement of the biophilic design approach, corresponding sustainable goals should be assigned for each pattern application, assuring that the patterns are not only satisfactory for human wellbeing but also responsible with mindfulness of their effect on earth resources.

REFERENCES

[1] Zhong, W., Schröder, T. & Bekkering, J., Biophilic design in architecture and its contributions to health, wellbeing, and sustainability: A critical review. *Frontiers of Architectural Research.* DOI: 10.1016/j.foar.2021.07.006.

[2] Browning, W.D., Ryan, C.O. & Clancy, J.O., Patterns of biophilic design. Terrapin Bright Green, LLC: New York, **14**, pp. 3–4.

[3] Gou., Z, Prasad, D. & Lau, S.S.-Y., Impacts of green certifications, ventilation and office types on occupant satisfaction with indoor environmental quality. *Architectural Science Review*, **57**(3), pp. 196–206, 2014. DOI: 10.1080/00038628.2014.908113.

[4] Kellert, S.R., *Nature by Design*, Yale University Press, 2018.

[5] Chang, C.-Y. & Chen. P.-K., Human response to window views and indoor plants in the workplace. *HortScience*, **40**(5), pp. 1354–1359, 2005.

[6] Fei, X., Lau, S.S.-Y., Gou, Z., Song, Y. & Jiang, B., Incorporating biophilia into green building rating tools for promoting health and wellbeing. *Environmental Impact Assessment Review*, **76**, pp. 98–112, 2019.

[7] Schweitzer, M., Gilpin, L. & Frampton, S., Healing spaces: Elements of environmental design that make an impact on health. *Journal of Alternative and Complementary Medicine*, **10**(1), p. S-71, 2004.

[8] Nugent, N. & Rhinard, M., *The European Commission*, Macmillan International Higher Education, 2015.

[9] United Nations, Resolution adopted by the General Assembly on 6 July 2017, Work of the Statistical Commission pertaining to the 2030 Agenda for Sustainable Development (A/RES/71/313 Archived 28 November 2020 at the Wayback Machine), 2017.

[10] Browning, W.D., Ryan, C.O., Clancy, J.O., *14 Patterns of Biophilic Design*, Terrapin Bright Green, LLC: New York, 2014.

RESIDENTIAL NEIGHBORHOOD ASSESSMENT IN THE CITY OF AL AIN, UNITED ARAB EMIRATES, AND THE IMPACT ON CLIMATE CHANGE (HEAT ISLAND EFFECT ANALYSIS)

LINDITA BANDE, MOUAZ MOHAMED, YOSAN ASMELASH & AFRA ALNUAIMI
[1]United Arab Emirates University, UAE

ABSTRACT

Al Ain is an important city that has gone through major urban transformation over the last 50 years. These changes impact the development of neighborhoods, especially residential neighborhoods. The morphology of the urban blocks also impacted the heat island effect (UHI). UHI is one of the contributors to climate change. Therefore, having sustainable neighborhoods' means reducing the UHI and making cities more livable. The methodology to be followed in this study is as residential blocks in Al Ain for analysis of current conditions and proposal for UHI reduction through different strategies, modelling and simulations to create and analyze the models in Rhino Grasshopper and discussing the findings and results. The aim is to divide the work for the two cities and then analyze the current conditions of the neighborhoods. Based on the findings, different strategies can be applied in the residential blocks to reduce the surface temperature of the streets and buildings, therefore reducing the UHI effect. Having sustainable neighborhoods has a direct impact in making cities more livable and improving the climate change. The main tools to be used are Rhino/Grasshopper. Through advanced software, the findings can be optimized and contribute to more sustainable neighborhoods.
Keywords: climate change, neighborhood assessment, grasshopper, walkability.

1 INTRODUCTION

Al Ain is one of the world's oldest inhabited communities, located at the crossroads of the ancient regions of the Arabian Peninsula and Mesopotamia (3,000 BC). The oasis around which a Bedouin tribe resided gave Al Ain its name, and the city became known as the United Arab Emirates' "Garden City" [1]. In barely over 30 years, Al Ain has transformed from a desert oasis to a thriving contemporary city [2]. Several reasons drew civilization to Al Ain throughout antiquity, including the availability of freshwater, date oases, low humidity, and Al Ain's function as a transit hub between interior areas and the Arabian Gulf [3]. Al Ain is noted for its infrastructure, which is a mix of contemporary and traditional. In contrast to the abundant towering skyscraper typology in Abu Dhabi and Dubai, Al Ain has instituted a policy of allowable maximum building height of four stories including the ground level, which was established by the country's founder, Sheik Zayed, to preserve the city's cultural quality [4]. When it comes to traffic patterns, Al Ain has a system of roundabouts that are now being converted into traffic signals. Al Ain is a favorite weekend and vacation destination for families from all around the UAE, especially Abu Dhabi. Al Ain has grown in popularity as a tourist destination, attracting people from around the Gulf Cooperation Council (GCC) [5]. The latest construction activity, which is distinguished by the city's mountain, Jebel Hafeet, consists of around 11,000 big Mediterranean-style houses with extensive green landscaping and an artificial canal [1]. Other higher education institutions, an international airport, and several tourist attractions are also located in the city.

WIT Transactions on The Built Environment, Vol 210, © 2022 WIT Press
www.witpress.com, ISSN 1743-3509 (on-line)
doi:10.2495/ARC220111

1.1 Planning of Al Ain

As they came into touch with Islamic culture, the Arab Gulf cities are considered as a fusion of many civilizations and eras, eventually resulting in the urban structures and morphological forms observed in current Arab cities. As a consequence of growing income and a move to an urban services-based economy, many Arab cities' urban design and settlement patterns have altered considerably. Climate and social-religious restrictions shaped traditional Arab urban form. The master-planning phase, which is normally carried out by western experts, is incompatible with the Gulf's economic, political, socio-cultural, and environmental dynamics. Traditional Arab communities took one of two forms: "planned" or "spontaneous" urban layouts.

The main objectives of the first General Master Plan of Al Ain city [17]. Continuation of the traditional role of Al Ain as a center for educational, cultural, and recreational activities of national importance, as the main agricultural producer of the Emirate:

1. Minimization of water consumption to ensure a reserve of groundwater;
2. Economic diversification to provide stable conditions for local businesses;
3. Agricultural self-sufficiency for products which Al Ain is suited;
4. Reducing work contracts and increasing employment for citizens, to support the sustainable economic development of the region;
5. Maintenance of high-level services.

1.2 Building typology of Al Ain

Al Ain contains a city center as well as additional commercial structures made up of mid-rise buildings that are four to five stories tall. The villa is the most common type of residential structure in Al Ain, and it is frequently found in a compound with a similar model organization with periodically or as individual units. It also has many districts with substandard tenement housing and the Al Ain municipality initiated a campaign that stands against unlicensed construction and building code violations [18].

1.3 Urban heat island effect

The goal of sustainable development is to create a blueprint that represents peace and prosperity. An issue city around the world faces is the urban heat island effect. The urban heat island effect is a type of heat buildup phenomenon caused by human activities and urban building construction in an urban area. This occurs when natural grounds are replaced with large concentrations of heat-absorbing buildings, pavement, and different surfaces. This increase causes a high increase in energy consumption cost, material flow, energy flow, illnesses due to heat, mortality, and air pollution. The urban heat island effect develops from solar radiation that is absorbed by the constructed surfaces of tall buildings, concrete structures, roofs, and roadways, which is then released as a form of heat. "Urban heat island" refers to built-up areas that have higher heat energy levels from the surrounding environment [10], [12]. It makes you more susceptible to respiratory and airborne illnesses. The demand for power rises when air conditioning is required to temper the heat, increasing the greenhouse effect. The need for water is likewise increasing [11].

1.4 Influence factors

Many factors have a prominent role in increasing UHI throughout cities. Minimum landscape or decreasing the amount of natural landscape in an urban area; planting/utilizing vegetation, water bodies, and trees provide shading which makes the area cooler. Additionally, Sidewalks, roofs, buildings, roads, and parking have a dry, solid surface in an urban area which results in low shade and moisture from natural landscaped areas which lead to an increase in temperature. The properties of materials also have a noticeable effect; when compared to trees, vegetation, and other natural surfaces, conventional materials that are human-made and are being used in urban contexts have low reflectivity and have a high emission of heat energy. Wind movement and the ability of urban materials to absorb and release solar energy are influenced by the proportions and spacing of structures inside a city. Surfaces and structures obscured by nearby buildings produce huge thermal masses that cannot easily release their heat in densely urbanized areas. Cities with a lot of small streets and towering buildings produce urban canyons, which can hinder natural wind flow and reduce cooling. Additionally, heat is emitted into the urban environment by vehicles, air conditioning systems, buildings, and industrial operations. Heat island effects can be exacerbated by these sources of human-generated, or anthropogenic, waste heat [13].

Finally, the weather is a critical element to consider in regards to UHI. By maximizing the quantity of solar radiation hitting urban surfaces and decreasing the amount of heat that can be moved away, calm and clear weather conditions result in more severe heat islands. Intense winds and cloud cover, on the other hand, prevent heat islands from forming. The heat island effect can be influenced by geographical factors as well. Nearby mountains, for example, can prevent wind from reaching a city or produce wind patterns that pass past it [13].

1.5 Effects on sustainability

The urban heat island effect can cause temperature rises in cities, which can have a variety of consequences for the three pillars of sustainability. Economy, people, and the environment are the three foundations. High energy consumption is a prominent result of UHI. As the temperature rises, the need for cooling rises as well, placing additional strain on the electrical grid. There is also a notable increase in emissions of greenhouse gases and air pollutants. As more electricity is required to cool the building's surfaces, the demand for energy rises, resulting in increased emissions of greenhouse gases and air pollutants from power plants. Regarding water demand, water usage is high both indoors and outdoors [8].

1.6 Neighborhood assessment

Bande et al. [15] investigated how to efficiently cut energy consumption for the villa typology, a significant building form for Abu Dhabi and many other towns in the area. The goal of this study is to use a calibrated model to highlight the effects of suggested energy efficiency improvements on a villa and to show that, owing to the importance of the urban heat island effect, the model must be run using urban weather data rather than rural weather data to be correct. Construction properties, on-site (urban) weather data, occupancy-related loads and schedules, and rural weather data are among the data sources for this case study. Weather data customization integrating urban and rural weather variables, model calibration using a genetic algorithm-based tool, and simulation of retrofit schemes were the four primary processes. The results reveal that the cool roof, cool walls, SSEER, and atrium shades are the best techniques for a villa. This study's recommended mix of solutions reveals that overall energy savings may be lowered by 25%. The authors recommend that more research be done to improve the software so that the techniques can work at their best [15].

Another study looked at temperature variations in desert city-regions [14]. In order to explore land cover and temperature relationships, the authors developed a multi-sensory strategy. The temperature trend was linked to NDVI and ISA percentage fluctuations in the research. Finally, in arid city locations, the inversion of the Urban Heat Island phenomena was highlighted. Due to the urban heat island effect, the center area of the city is typically anticipated to have higher temperatures. This effect, however, is commonly inverted in desert conditions, as demonstrated in this study. The comparatively high amount of vegetation in urban regions compared to suburban areas, where bare land and sand are more plentiful, is one of the key causes for this inversion [14].

Another paper attempts to investigate the amplitude of UHI and its relationship with the key meteorological factors (temperature, wind velocity, and wind direction) in the context of Dubai, United Arab Emirates. A clustering approach was used to post-process five years of hourly weather data ranging from 2014 to 2018 collected from weather stations in an urban, suburban, and rural region. Six clusters were studied, each with a different range of wind directions. According to the findings, UHI is influenced by synoptic weather conditions (e.g., sea breeze and warm air from the desert) and is greater at night. It is also emphasized that air temperature and evening Urban heat island intensity in the urban region are 1.3°C and 3.3°C higher than in the rural area, respectively, throughout a five-year period, and the Urban heat island effect and air temperature are only independent of each other when the wind originates from the desert [16].

2 METHODOLOGY

The research follows a linear methodology. The study starts with the neighborhood assessment, then weather analysis, modelling and simulations. The aim is to understand the district physiognomy and the connection to the surface temperature, solar radiation, outdoor thermal comfort (OTC).

2.1 Case study selection

The selected case study are 6 districts located int eh city of Al Ain. The typology is low rise and mid-rise Buildings. Table 1 shows the main areas, with the plan and site images. Table 2 shows more data on the area, Breen area available and shading devices if there are any. The six main districts are: Asharej: Mriefa compound complex, City Centre: Al Radi Street, AL Jimi: Jebel Hafeet complex, AL Jimi: Al Mada residential complex, Al Mu'tarid: Hazza bin Zayed stadium Area, Jebel Hafeet: New look medical center.

2.2 Al Ain weather conditions

The climate in Al Ain is considered a desert climate with minimum rainfall. Based on the latest climate analysis graphs of the weather in Al Ain using the graphics-based computer program Climate Consultant, Al Ain has an average temperature of 28°C which is considered above the comfort level. The favorable wind comes from north to west which carries a cool lightweight wind with a breeze, and the unfavorable wind comes from the north side which carries dust and has a hot breeze. Al Ain has prominent levels of solar radiation and illumination but the humidity level is low which results in low levels of rainfall. Al Ain has

Table 1: Selected districts.

District	Neighborhood	Plan	Site images
Asharej	Mriefa compound complex		
City Centre	Al Radi Street		
AL Jimi	Jebel Hafeet complex		
AL Jimi	Al Mada residential complex		
Al Mu'tarid	Hazza bin Zayed stadium area		
Jebel Hafeet	New look medical center	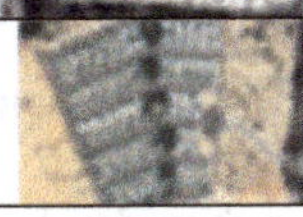	

Table 2: District details.

District	Building typology	District area	Green area/ number of trees	Shading furniture
Mriefa compound complex	Low rise	58,105.34 m^2	11	n/a
Al Radi Street	Low rise	91,480.27 m^2	46	n/a
Jebel Hafeet complex	Low rise	4.52 km^2	10	n/a
Al Mada residential complex	Mid-rise	24,062.25 m^2	10	n/a
Hazza bin Zayed stadium Area	Mid-rise	187,645.1 m^2	120	Three urban shading devices
New look medical center	Mid-rise	249,259.45 m^2	65	n/a

a unique microclimate in contrast to the larger, more humid coastal cities in the UAE. Despite its location in a hot and arid climate, the irrigated greenery proves to be a viable source of evapotranspiration and humidity (Figs 1 and 2) [7].

2.3 Modelling and simulation, tools and resources

Rhinoceros 3D-Rhinoceros, often known as Rhino or Rhino3D, is a CAD program created by Robert McNeel and Associates that allows users to explore a variety of three-dimensional

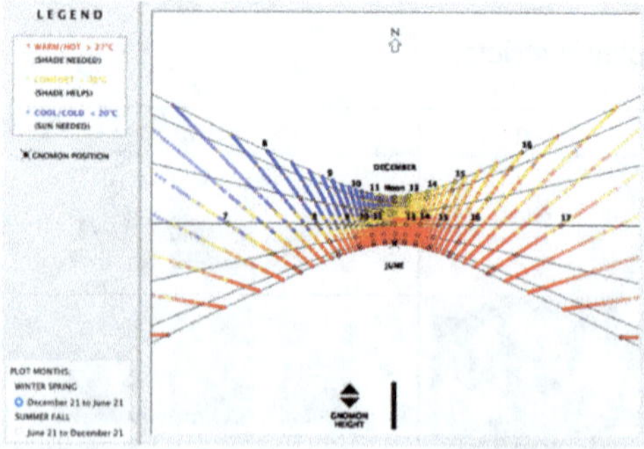

Figure 1: Temperature range graphs: ASHRAE Standard 55-2004 using PMV.

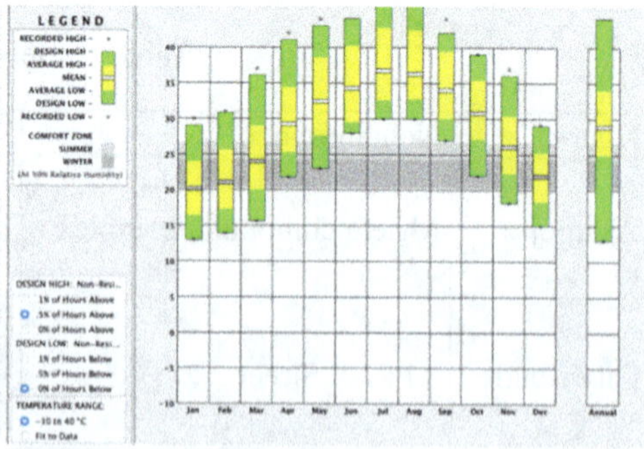

Figure 2: Temperature range graphs: ASHRAE Standard 55-2004 using PMV.

capabilities such as volume creation and design. Non-uniform rational B-splines, or NURBS, are used by Rhino3D to allow users to actualize concepts and designs.

Grasshopper: The Rhinoceros 3D application uses the Grasshopper visual programming language. Grasshopper is primarily used to develop generative algorithms in the shape of boxes and arrows; boxes are viewed as entities, with arrows representing relationships connecting them. Grasshopper, like its parent application Rhinoceros 3D, is a very adaptable tool with several plug-ins that may be used for a variety of applications.

Ladybug: In Grasshopper, Ladybug displays and analyses weather data. This comprises geometric studies such as radiation analysis, shade studies, and view analysis, as well as diagrams such as the sun route, wind rose, psychrometric chart, and so on.

Honeybee: Honeybee connects Grasshopper3D to proven simulation engines, such as EnergyPlus/OpenStudio (for energy performance, HVAC sizing, thermal comfort, and so on) and Radiance (for efficiency measures, HVAC sizing, thermal comfort, daylighting, and glare simulation, Fig. 3)

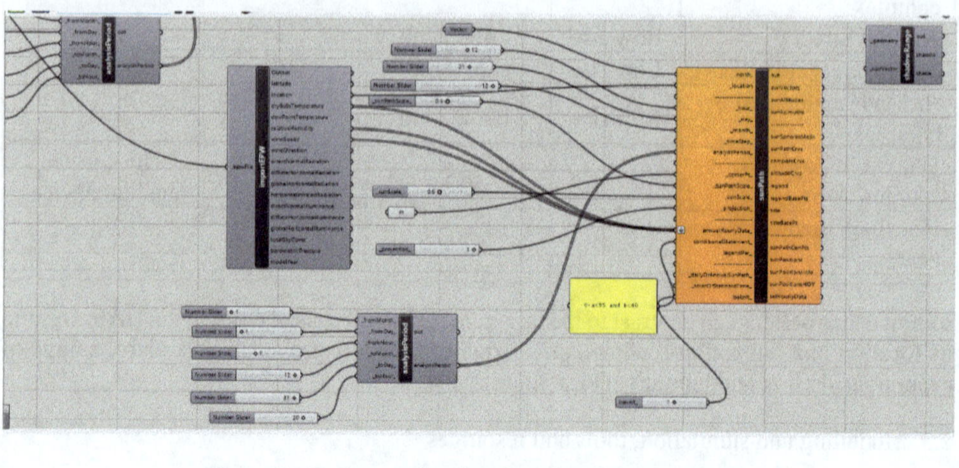

Figure 3: Python script for Grasshopper application.

2.4 Challenges faced during simulation

Both analyses were conducted in grasshopper using the ladybug plugin. the scripts were obtained from Hydrashare GitHub, a website where script creators share scripts for several types of analysis. Some of the models were manually modeled while the others were obtained from a purchased 3DS file for Al Ain city. There were several faults in the file as the 3D models were mostly for non-residential buildings, and those for the residential areas were not accurate for some cases and had to be remodeled. Moreover, it was difficult to locate neighborhoods and select specific buildings in the model, since it did not specify locations or have any labels. Running the scripts was a volatile process, and sometimes it ran promptly other times it took a few hours. Overall, the speed of the script depended on the GPU strength of the PCs used. To conclude, grasshopper could obtain the different results presented in this paper to be used in the research.

3 RESULTS

The results of the simulations refer to the solar radiation of the streets surface analyzed in the base case scenario and the modified scenario where the shades are inserted in the main streets and inner space of districts. Table 3 refers to the solar radiation in equinox and solstice days. Table 4 refers to the solar radiation in the six districts on a yearly analysis. Table 5 refers to the results of the UTCI of the basic case (large-scale) of the districts. Fig. 4 is a graphic representation of the solar radiation in one of the districts, Hazza Stadium area.

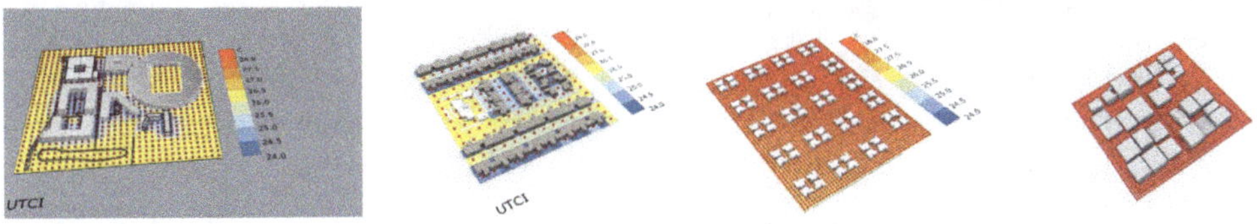

Figure 4: UTCI results.

4 DISCUSSION

The main focus of this study was to understand the impact of the district design and height of the buildings on the solar radiation and outdoor thermal comfort. Through the application of shading structure, the biggest impact is shown in low rise districts. Conducting this study was challenging due to the lack of available free data. Also collecting material on the base case was challenging.

The software used was Rhino/Grasshopper. The modelling and analysis of each case was proven to be challenging due to the complexity of the python language. Furthermore, the modelling phase needed more efforts due to the lack of data. Also parts of the software were customized as per each case.

However, the results are promising in understanding the relevance in designing sustainable neighborhoods in order to fight the climate change, improve walkability through reducing the heat island effect and improving the outdoor thermal comfort.

5 CONCLUSION

The aim is to divide the work for the two cities and then analyze the current conditions of the neighborhoods. Based on the findings different strategies can be applied in the residential

Table 3: Solar radiation analysis results.

Name	September 21	December 21	March 21	June 21	Annual data
HBZ stadium neighborhood				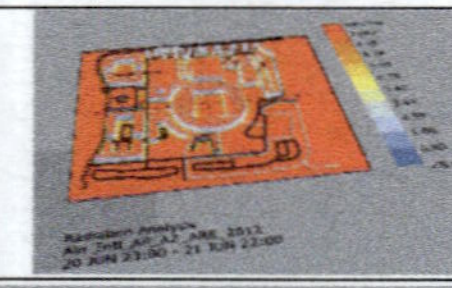	
New look clinic		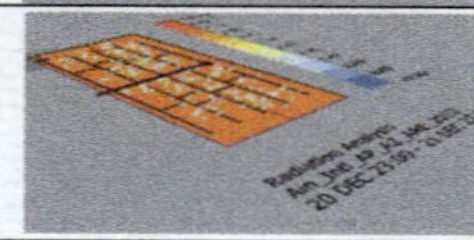	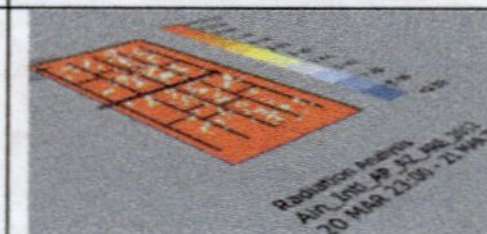		
Mriefa compound complex		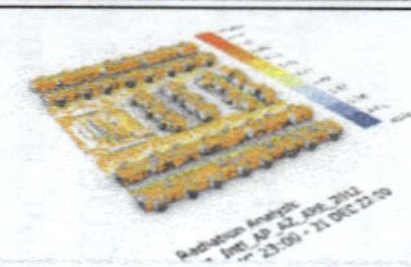	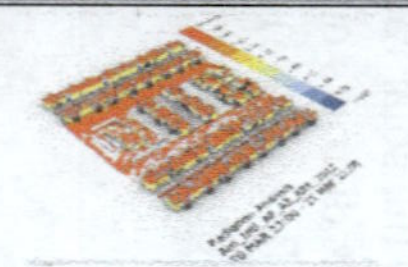	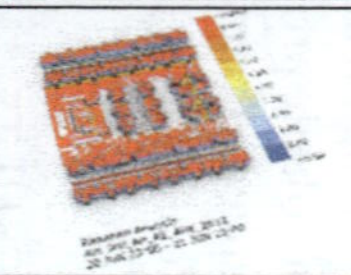	
City center Medclinic	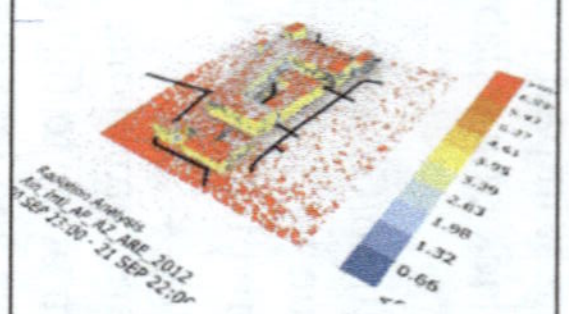				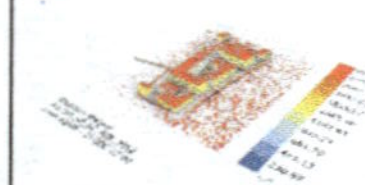
Al Radi Street		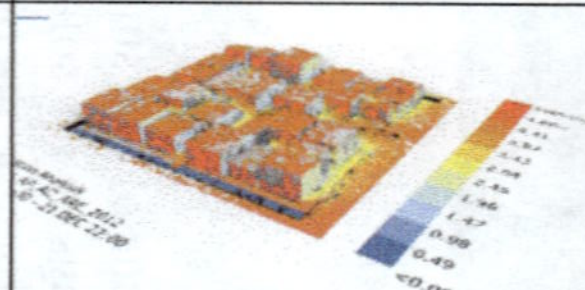	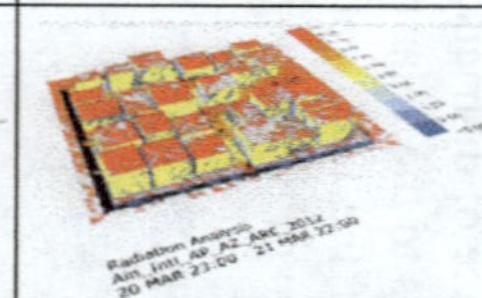	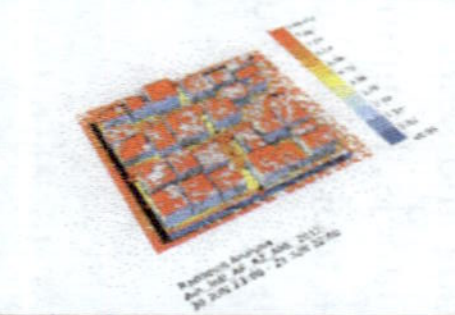	
Jebel Hafeet	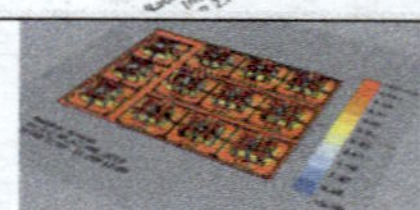	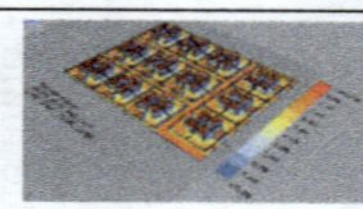		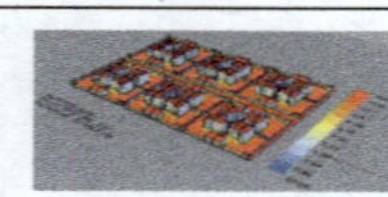	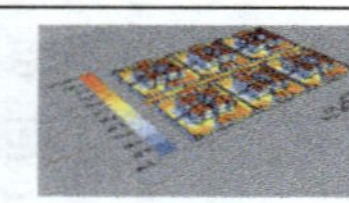

Table 4: Solar radiation results.

	Mriefa compound complex		Jebel Hafeet complex		Al Radi Street		New look medical center		Hazza bin Zayed stadium		Al Mada residential complex	
	No T.	With T.	No T.	With T.	No T.	With T.	No T.	With T.	No T.	With T.	No T.	With T.
January	63.9	58.3	53.4	44.0	44.1	38.2	52.3	50.1	50.9	32.7	59.1	52.6
February	66.9	62.1	58.6	49.2	46.7	40.8	55.2	52.8	53.9	33.5	62.5	56.3
March	83.3	78.3	77.1	65.3	59.3	52.5	69.1	66.1	67.7	40.6	78.7	71.6
April	98.6	85.0	87.5	75.1	64.7	57.7	74.7	71.2	72.3	41.8	84.8	77.6
May	63.9	78.3	96.8	84.3	70.8	63.4	81.6	77.6	78.4	44.6	92.1	84.4
June	95.1	93.3	92.4	80.6	68.4	61.2	71.2	74.9	75.5	42.8	88.7	81.2
July	97.3	86.7	95.2	83.1	69.9	62.6	80.6	76.6	77.2	43.8	90.8	83.2
August	93.6	92.2	91.7	79.1	67.1	60.0	77.5	73.7	74.6	42.7	87.6	80.3
September	83.4	88.5	78.2	66.8	59.7	52.9	69.2	66.2	67.4	39.8	78.7	71.7
October	77.2	78.7	78.2	58.0	54.2	47.3	63.7	61.0	62.3	38.2	72.2	65.2
November	63.6	58.4	53.6	44.4	44.2	38.2	52.1	49.9	50.8	32.3	59.0	52.6
December	60.3	57.4	49.7	40.8	41.6	36.0	49.4	47.3	47.9	31.0	55.9	49.5
Annual	947.6	918.0	932.1	796.0	691.2	630.7	797.4	792.3	779.5	478.9	939.4	852.8

Table 5: UTCI values.

	Mriefa compound complex		Jebel Hafeet complex		Al Radi Street		New look medical center		Hazza bin Zayed stadium		Al Mada residential complex	
	No T.	With T.	No T.	With T.	No T.	With T.	No T.	With T.	No T.	With T.	No T.	With T.
January	23.3	15.2	23.8	23.2	22.9	22.5	22.7	22.6	23.1	15.2	23.2	22.7
February	23.1	15.9	23.7	23.0	22.7	22.3	22.5	22.4	22.9	15.9	23.	22.6
March	25.9	20.2	26.4	25.8	25.6	25.2	25.3	25.3	25.7	20.2	25.9	25.4
April	30.7	27.0	31.2	30.7	30.4	30.1	30.2	30.2	30.6	27.0	30.7	30.3
May	34.2	31.6	34.6	34.1	33.9	33.5	33.7	33.7	34.0	31.6	34.1	33.8
June	36.3	34.2	36.7	36.2	36.0	35.7	35.9	35.8	36.2	34.2	36.3	35.9
July	39.1	37.5	39.5	39.0	38.8	38.5	38.7	38.6	39.0	37.5	39.1	38.7
August	38.2	36.6	38.7	38.2	38.0	37.7	37.8	37.8	38.1	36.6	38.2	37.9
September	35.6	33.3	36.1	35.6	35.4	35.0	35.2	35.1	35.5	33.3	35.6	35.2
October	32.3	28.9	32.8	32.2	32.0	31.7	31.8	31.8	32.2	28.9	32.3	31.9
November	27.1	21.5	27.6	27.0	26.8	26.4	26.6	26.5	26.9	21.5	27.1	26.6
December	24.2	17.3	24.7	24.1	23.8	23.4	23.6	23.5	24.0	21.5	24.1	23.7
Annual	30.8	26.6	30.8	30.2	30.0	29.6	30.3	30.3	30.1	27.0	30.3	29.9

blocks in order to reduce the surface temperature of the streets and buildings, therefore reduce the UHI effect. Having sustainable neighborhoods has a direct impact in making cities more livable and improving the climate change. The main tools to be used are rhino/grasshopper. Through advanced software, the findings can be optimized and contribute to more sustainable neighborhoods.

Future work is needed to investigate other means of reducing the solar radiation in the streets of the city. This study can contribute to improving the OTC, reducing heat island effect in the city and the UAE.

REFERENCES

[1] El Amrousi, M., Paleologos, E., Caratelli, P. & Elhakeem, M., Are garden cities in the desert sustainable? *International Review for Spatial Planning and Sustainable Development*, **6A**(1), pp. 79–94, 2018.

[2] Haggag, M. & Hadjri, K., The transformation of residential patterns in Al-Ain city, UAE. *WIT Transactions on Ecology and the Environment*, vol. 84, WIT Press: Southampton and Boston, 1970.

[3] Yagoub, M., Monitoring of urban growth of a desert city through remote sensing: Al-Ain, UAE, between 1976 and 2000. *International Journal of Remote Sensing*, 25(6), pp. 1063–1076, 2004.

[4] Hamdan, D. & de Oliveira, F., The impact of urban design elements on microclimate in hot arid climatic conditions: Al Ain City, UAE. *Energy and Buildings*, **200**, pp. 86–103, 2019.

[5] Eid, R. & Elbanna, S., A triangulation study to assess the perceived city image in the Arab Middle East context: The case of Al-Ain in the UAE. *Tourism Planning and Development*, **15**(2), pp. 118–133, 2017.

[6] Climate-data.org, Al-Ain climate: Average temperature, weather by month, 2022. https://en.climate-data.org/asia/united-arab-emirates/abu-dhabi/al-ain-1210/. Accessed on: 13 Jan. 2022.

[7] Bande, L., Manandhar, P., Ghazal, R. & Marpu, P., Characterization of local climate zones using ENVI-met and site data in the city of Al-Ain, UAE. *International Journal of Sustainable Development and Planning*, **15**(5), pp. 751–760, 2020.

[8] Swain, S., The impact of urban heat island effect on sustainability. EcoMENA, 2022. https://www.ecomena.org/uhi-effect-sustainability/. Accessed on: 13 Feb. 2022.

[9] Abulibdeh, A., Analysis of urban heat island characteristics and mitigation strategies for eight arid and semi-arid gulf region cities, 2022.

[10] UNEP, Urban heat island: Its effects in UAE, Ambassador report – Our actions. 2022. https://tunza.eco-generation.org/ambassadorReportView.jsp?viewID=43331. Accessed on: 13 Feb. 2022.

[11] UNEP, Urban heat island: Its effects in UAE, Ambassador report – Our actions. 2022. https://tunza.eco-generation.org/ambassadorReportView.jsp?viewID=43331. Accessed on: 13 Feb. 2022.

[12] Al Blooshi, L., Abuelgasim, A., Nassar, A. & Ksiksi, T., Scirp.org, 2022. https://www.scirp.org/pdf/ojg_2020072417101026.pdf. Accessed on: 13 Feb. 2022.

[13] US EPA, Reduce urban heat island effect, 2022. https://www.epa.gov/green-infrastructure/reduce-urban-heat-island-effect. Accessed on: 13 Feb. 2022.

[14] Lazzarini, M., Marpu, P. & Ghedira, H., Temperature-land cover interactions: The inversion of urban heat island phenomenon in desert city areas, 2022.

[15] Bande, L., Cabrera, A., Kim, Y., Afshari, A., Ragusini, M. & Cooke, M., A building retrofit and sensitivity analysis in an automatically calibrated model considering the urban heat island effect in Abu Dhabi, UAE, 2019.
[16] Mohammed, A., Pignatta, G., Topriska, E. & Santamouris, M., Canopy urban heat island and its association with climate conditions in Dubai, UAE, 2022.
[17] Al-Ain Municipality, General Master Plan of Al Ain city. 1st ed., Town Planning Department: Al-Ain, 1886.
[18] Dmt.gov.ae, *Home*, 2022. https://www.dmt.gov.ae/en/aam. Accessed on: 13 Feb. 2022.

SECTION 5
INNOVATIVE MATERIALS AND PRODUCTS

DURABILITY PERFORMANCES OF COMPRESSED EARTH BLOCKS EXPOSED TO WETTING–DRYING CYCLES AND HIGH TEMPERATURE

PHILBERT NSHIMIYIMANA[1], CESAIRE HEMA[1], SEICK OMAR SORE[1,2], OUSMANE ZOUNGRANA[1], ADAMAH MESSAN[1] & LUC COURARD[3]
[1]Laboratoire Eco-Matériaux et Habitats Durables, Institut International d'Ingénierie de l'Eau et de l'Environnement, Burkina Faso
[2]Département Génie Civil de l'Institut Universitaire de Technologie, Université Nazi BONI, Burkina Faso
[3]Urban and Environmental Engineering, Université de Liège, Belgium

ABSTRACT

The environmental durability of earthen materials is among the main factors that limit their widespread acceptance in the contemporary building sector. This study assesses the performances of compressed earth blocks (CEBs) before and after exposure to the wetting–drying (WD) cycles and high temperature. The CEBs were produced from kaolinite-rich earth material stabilized with lime-rich residue (5%–25%) or the lime substituted with rice husk ash, lime:ash (20:0%–12:8%) or cement (4%–8%), with respect to the mass of dry earth. The CEBs were cured for 28–45 days and at the ambient temperature of a laboratory of 30±5°C and the moisture of production, necessary for the reactivity of the binders and improvement of the performances of CEBs. The CEBs were dried before testing their initial compressive strength. Their compressive strength was also tested after exposure to 12 cycles of wetting in water for 6 hours at 30±5°C and drying in the oven for 42 hours at 70±5°C. Additionally, it was tested after exposure to the elevated temperature of 150–600°C. After exposure to the WD cycles, the compressive strength of CEBs relatively increased, by up to 49% (4.6 to 6.8 MPa) for CEBs stabilized with lime (15%) and by up to 40% (4.4 to 6.2 MPa) with lime:ash (20:0%), with respect to their strength before the WD cycles. The maximum increase of the strength was observed for CEBs containing a higher amount of lime, related to the reaction of excess lime which resulted in the formation of more cementitious products and improvement of the strength of CEBs. Nevertheless, the compressive strength decreased by 55% (12.5 to 5.5 MPa) for the CEBs stabilized with cement (4%), resulting from the degradation of the initial cementitious products. However, the strength increased by up to 58% (4.2 to 6.6 MPa) for the CEBs stabilized with cement (8%) after exposure to 600°C. This implies that the stabilization of CEBs with lime-rich binder is more resilient to the WD cycles than cement. It also shows that the cement stabilized CEBs would at least retain their strength after exposure to high temperature.

Keywords: compressed earth blocks, chemical binder, compressive strength, high temperature, pozzolanic reaction, wetting–drying cycles.

1 INTRODUCTION

The improvement of the durability performances of earthen materials, and more specifically compressed earth blocks (CEBs) is essential for their widespread adoption in the building construction [1], [2]. Numbers of studies have reported that the different stabilisation using chemical industrial binders: cement and lime or alternative/by-product binders: geopolymer, lime residue or riche husk ash improve various engineering and durability indicators of CEBs [3]–[7].

Indeed, a very recent literature review has highlighted very interesting points on various advances currently made on the improvement of performance of stabilized CEBs [8]. The review also highlighted numbers of limitations and missing parameters of CEBs that still need to be addressed or tested, such as the lack of data on the "sound insulation and fire resistance properties", among other parameters. It was discussed that the performances of

WIT Transactions on The Built Environment, Vol 210, © 2022 WIT Press
www.witpress.com, ISSN 1743-3509 (on-line)
doi:10.2495/ARC220121

stabilized CEBs not only depends of the quality and the reactivity of the selected earthen materials, but also the conditions of production/ curing and testing [9], [10]. This suggests that the conditions of exposure would also affect the performances of CEBs.

In fact, few studies have investigated the behaviour of earthen materials after exposure to high temperature (fire). Beckett and Kazamias [11] were only interested in predicting the evolution of thermal behaviour of unstabilised rammed. Abdallah et al. [12] reported that the water content affects the thermal instability, while the cement stabilisation improves the thermal stability of earth. One study reported that increasing the firing temperature increases the compressive strength and reduces the abrasion coefficient of stabilised CEBs [13]. Moreover, some studies have shown that the CEBs stabilized with cement loss their strength after exposure to the wetting–drying (WD) cycles. The compressive strength decreased, from the value before to the value after the WD cycles, 0.3 and 0.5 times respectively when the CEBs were stabilized with 4% and with 8% cement [14], [15]. However, these studies did not assess the effects of exposure conditions on the engineering and durability performances of earthen materials, specifically the CEBs stabilized with by-product binders.

The present study reports on the preliminary results on the assessment of the performances of stabilized CEBs, when they are exposed to extreme conditions such WD cycles and high temperature. The current results are based on the evolution of the mechanical properties: the compressive strength of stabilized CEBs before and after exposure to such conditions. The present study specifically aimed to highlight the differences in the behaviours of stabilized CEBs depending on the type of chemical stabilizers, i.e. hydraulic binders: cement and lime versus pozzolanic binders: rice husk ash.

2 MATERIALS AND METHODS

2.1 Materials and methods for the production of CEBs

The present study was carried out using materials collected locally in the vicinity of Ouagadougou, Burkina Faso. The earthen material is a kaolinite-rich clay (60%) material collected from the site of Kamboinse (12°29'24.48" N; 1°32'59.28" W). The previous study have reported its physical, chemical and mineralogical properties [16]. The earthen material was stabilized with by-product binders for the production of stabilised CEBs. The lime residue (calcium carbide residue), collected from an acetylene production industry in Kossodo (12°25'56.1" N, 1°29'22.44" W). It was also stabilised using rice husk ash obtained by the calcination (500° for 2h) of the rich husk collected from Bagré (11°28'25.8" N, 2°03'21" W). The lime (residue) and the (riche husk) ash contain mainly hydrated lime ($Ca(OH)_2$) and amorphous silica, respectively, as it was reported in a previous study [17]. Moreover, the earthen material was stabilised with cement, CEM II 42.5 produced locally.

The CEBs were produced by mixing the earthen materials stabilized with 0 to 25% lime of the dry mass of the earthen material, 20:0 to 12:8% lime:ash and 4 to 8% cement. The dry mixtures were humidified with appropriate amount of water: 17 to 23% with lime or lime: ash, and 19% with cement, corresponding to the optimum moisture content of each mix. The humid mixtures were manually compressed in Terstaram press machine to produce stabilized CEBs: $29.5 \times 14 \times 9.5$ cm^3 [18]. The curing was carried out in ambient conditions, at the production moisture, for 45 days and 28 days respectively for CEBs stabilised with lime or lime:ash [10] and cement [18]. The cured CEBs were dried to constant mass [5], before the treatment by exposure to various conditions.

2.2 Methods for the treatment and characterization of CEBs

The cured and dried CEBs were exposed to 12 cycles of wetting–drying. One cycle of WD is equivalent to 6 hours soaking in water (at ambient temperature: 30±5°C), followed by 42 hours drying in oven at 70±5°C. Afterward, the samples were dried to constant mass before testing their performances. The cured and dried CEBs were also exposed to higher temperature, in the range of 150–600°C, at the step of 150°C, by heating in a muffle furnace which has a maximum heating capacity of 3000°C. For all temperature of exposure, the heating rate was set at 1°C/min. A heating rate of 10°C/min was also applied to quickly reach the 600°C, in order to assess effect of higher heating rate on the performances of CEBs. The samples were maintained at the soaking temperature for 2 hours and cooled naturally in the furnace, before their characterization.

The CEBs were characterized based on their mechanical performances, tested before and after exposure to the WD cycles or the high temperature. The compressive strength was tested referring to the standard PR XP P13-901 [19], using hydraulic press equipped with a 300 kN capacity load cell, at loading rate of 0.2 mm/s. The compressive strength, Rc (MPa), was calculated using eqn (1); where Fr (kN) is the maximum load at failure and S (cm^2) is the area of loaded surface:

$$\mathrm{Rc} = 10 \times \mathrm{Fr}/\mathrm{S}. \tag{1}$$

3 RESULTS AND DISCUSSION

3.1 Effect of exposure to the physical stability

The physical stability of the CEBs was affected by the exposure to both the wetting–drying (WD) cycles and high temperature. Fig. 1(a) shows that the CEBs were intact before the WD cycles. Fig. 1(b) shows the physical deterioration by the loss of particles on the edges of sample after exposure to the 12 cycles of WD, especially for the CEBs stabilized with low content of lime (5%). This is due to the weakening of the cohesion between particles. On the contrary, the CEBs stabilized with high content of lime (20%) maintained their physical stability after exposure to the 12 cycles of WD (Fig. 1(c)). Moreover, Fig. 2 did not show major physical change before and after exposure to high temperature, except the occurrence of radial crack on the surface of the CEBs (Fig. 2(c)). This may be related to the thermal shock felt in the CEBs due to the thermal gradient between the outside and the inside of the samples and/or the thermal shrinkage.

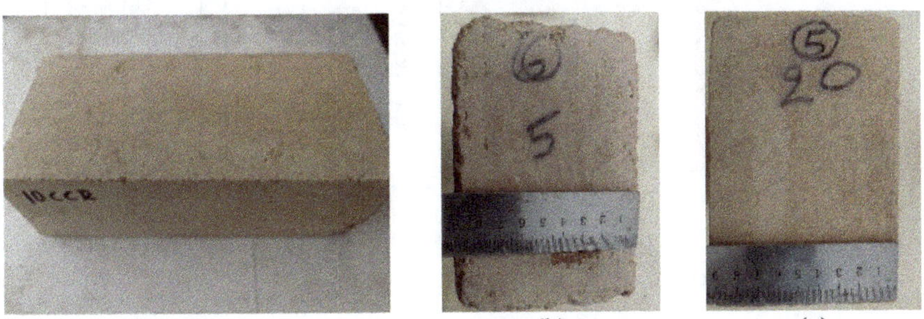

(a) (b) (c)

Figure 1: Samples of CEBs. (a) Before; vs (b) and (c) After 12 WD cycles, stabilized with 5% and 20% lime residue: deterioration of edges by the “loss of particle”.

Figure 2: Samples in the muffle furnace. (a) Before; (b) After exposure to high temperature; and (c) Surface of samples after exposure to 600°C: occurrence of the "radial cracks" on surface.

3.2 Effect of exposure to the wetting–drying cycles on the mechanical resistance

The mechanical performance of the CEBs stabilized with lime-rich residue (5%–25%) increases, before and after the 12 cycles of WD, reaching the maximum value up to 10%–15% of lime content. The unstabilised CEB did not withstand the exposure to the WD cycles. The compressive strength increased from 3 to 4.6 MPa before the WD cycles and 3.5 to 6.8 MPa after the WD cycles for the CEBs stabilized with 5% to 15% lime (Fig. 3(a)). Fig. 3(a) shows that, for the same content of the lime, the compressive strength of stabilized CEBs improves with the WD cycles. This improvement is relatively as low as 17% (3 to 3.5 MPa) for lower content of lime (5%), contrary to the improvement of up to 49% (4.6 to 6.8 MPa) recorded for higher content of lime (15%). Fig. 3(b) further presents the values of the relative compressive strength: the ratio between the values after the WD cycles and before the WD cycles. It shows that the values of the compressive strength after exposure to the WD cycles are significantly higher (up to 1.5 times) than the values before the WD cycles.

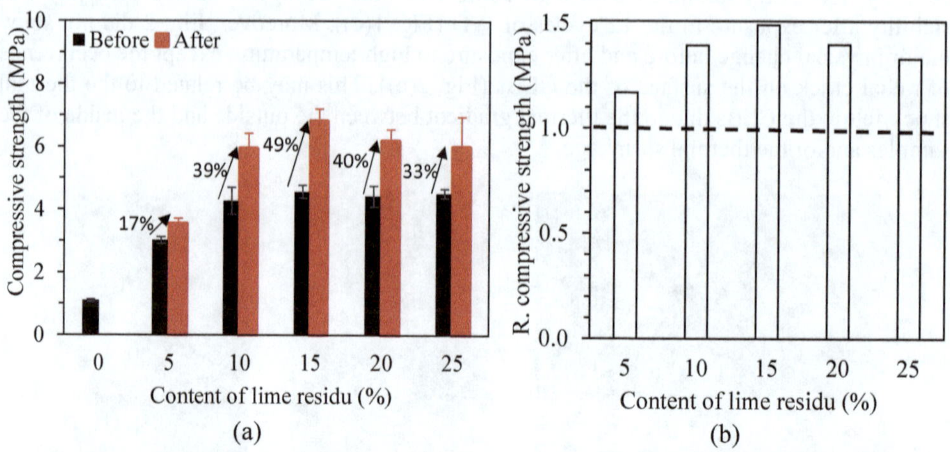

Figure 3: Compressive strength of CEB stabilized with lime. (a) Before and after 12 cycles of wetting–drying; and (b) Relative compressive strength: ratio between the values before and after WD cycles.

Fig. 4 presents the evolution of the compressive strength of CEBs stabilized with the mixtures of lime residue and rice husk ash in various ratios (lime:ash). It shows that the compressive strength increases with the ratio of lime:ash from 20:0 to 14:6 and decrease beyond, before and after the 12 cycles of WD. The average values of the compressive strength increased from 4.4 MPa to the maximum value of 8.4 MPa before the WD cycles and 6.2 MPa to the maximum value of 9.5 MPa after the WD cycles (Fig. 4(a)). The CEBs stabilized with 20:0% lime recorded the highest increase of the strength up to 40% (4.4 to 6.2 MPa) after the WD cycles. This shows that the compressive strength of CEBs is rather improved through the exposure the wetting drying cycles, as further presented in Fig. 4(b) where all the values of the relative compressive strength are greater than 1.

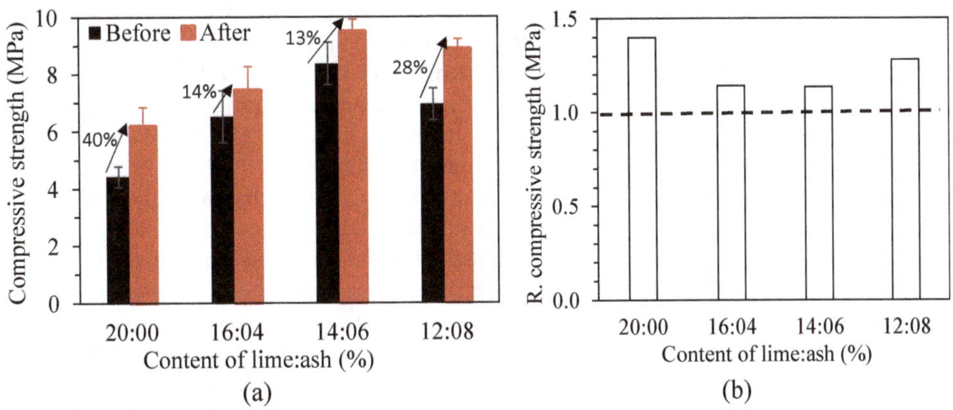

Figure 4: Compressive strength of CEB stabilized with lime:ash. (a) Before and after the 12 cycles of wetting–drying; and (b) Relative compressive strength.

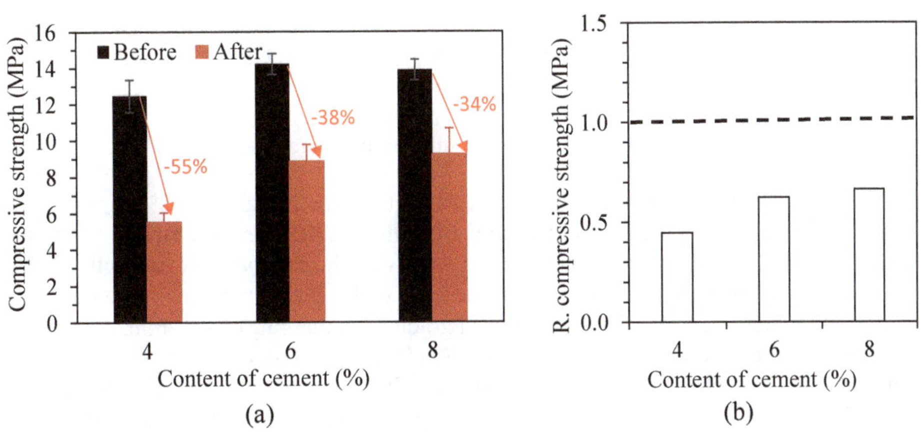

Figure 5: Compressive strength of CEBs stabilized with cement. (a) Before and after the 12 cycles of wetting drying of CEB; and (b) Relative compressive strength.

On the contrary, the compressive strength decreases for the CEBs stabilised with cement, as shown in Fig. 5. Fig. 5(a) shows that the average value of the compressive strength decreases by 55% from 12.5 to 5.5 MPa, 38% from 14.2 to 8.9 MPa and 34% from 13.9 to 9.2 MPa respectively for CEBs stabilised with 4%, 6% and 8% cement, before and after the WD cycles. It confirms that the exposure to the WD cycles degrades the mechanical performances of CEBs. This is further presented in Fig. 5(b), where the relative compressive strength remained far below 1 with all content of stabilization.

The increase of the compressive strength with the WD cycles, especially for CEBs stabilised with higher content of lime (≥ 15%) can be related to the effect of the treatment temperature (70°C) which, in the presence of the humidity, resulted in more reactivity in the CEBs matrix. This is also in agreement with the physical stability observed for the CEBs stabilized with 20% lime contrary to those stabilized with 5% clime, after exposure to the WD cycles (Fig. 1). In fact, the increase of the compressive strength of CEBs stabilized with the lime residue was previously related to the pozzolanic reaction between the earthen materials and the hydrated lime ($Ca(OH)_2$) contained in the residue [20]. At ambient temperature of curing, this reaction reaches the optimum with 10%–15% lime and so does the compressive strength. It was also reported that the kinetics of this reaction increase with increasing the curing temperature from 20°C to 40°C. It resulted in the increase of the compressive strength by 70% (2.3–3.9 MPa) and by 88% (2.5–4.7 MPa) respectively for CEBs stabilized with 10% and 20% lime residue [10]. This shows that the higher the content of lime, the more the effect of the temperature on the curing, and so does on the post-curing treatment. In the present study, the WD treatment, after the curing period, allowed for further reaction of the excess lime which did not successfully react during the curing and the increase of the compressive strength (Fig. 3). The same explanation can be made on the CEBs stabilized with the mixture of lime:ash, where the increase of the compressive strength was rather lower with the substitution of the lime residue by the rice husk ash (Fig. 4).

However, a different effect occurred for the CEBs stabilized with cement, which recorded a significant degradation of the compressive strength with the WD cycles (Fig. 5). Similar results were previously reported that the compressive strength of CEBs stabilized 4% and 8% cement respectively decreases by 0.3 times after six cycles of WD and 0.5 times after the 12 cycles of WD [14], [15]. The decrease of the compressive strength can be explained by the loss of the hydrates of calcium silicates and aluminates, previously formed during the curing of the stabilized CEBs. This results in the degradation of the cohesion in the matrix of CEBs and the loss of the strength.

3.3 Effect of exposure the high temperature on the compressive strength

Fig. 6 presents the evolution of the compressive strength of CEBs stabilised 8% cement, after exposure to the relatively higher temperature: from the ambient temperature of laboratory (30±5°C) to 600°C. It shows that the average values of the compressive strength did not significantly change from the ambient temperature up to the exposure temperature of 450°C, beyond which it increased. At 600°C, reached at gradual heating rate of 1°C/min, the average value of the compressive strength increased by 58% (4.2–6.6 MPa). Comparable results were recorded at 600°C, reached at rapid heating rate of 10°C/min, where the value of the strength increased by 46% (4.2–6.1 MPa) (Fig. 6(a)). It suggests that the CEBs stabilised with cement can at least maintain their strength up to 600°C. This is clearly shown in Fig. 6(b), where the relative compressive strength is around 1 up to 450°C and greater than 1 at 600°C. Similar observation is made on CEBs after exposure to 600°C reached at accelerated kinetics (10°C/min).

The strength of CEBs stabilized with cement after exposure to the temperature as high as 600°C, at both gradual and rapid increase is promising. This allows to predict that these CEBs

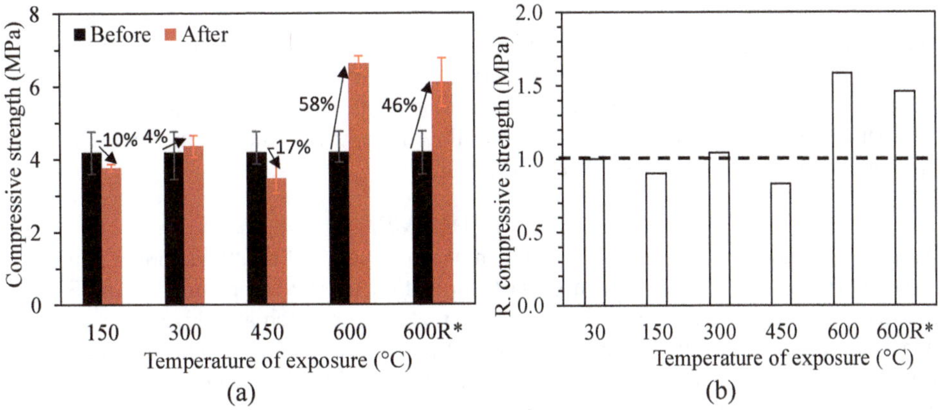

Figure 6: Effect of the temperature of exposure on: (a) The compressive strength; and (b) Relative compressive strength of CEB stabilized with 8% cement.

would potentially resist to the exposure at elevated temperature, such as in the case of fire, in agreement with [13]. On the contrary, the concrete would lose its strength and physical stability at high temperature (400–600°C) [21], [22]. It is also noteworthy highlighting the mechanical stability of CEBs in spite of their physical instability: occurrence of cracks on the surface of CEBs exposed to high temperature. However, these cracks may affect the transfer properties: porosity, water/vapor transfer in the CEBs. Therefore, they need further investigation.

4 CONCLUSIONS AND RECOMMENDATIONS

The present study assessed the performance of stabilized CEBs based on the evolution of their compressive strength before and after exposure to the wetting–drying cycles and high temperature. The following conclusions can be drawn:

- The compressive strength increases from 3 to 4.6 MPa for CEBs stabilized with 5% to 25% lime, before the WD cycles. It further increases from 3.5 to 6.8 MPa after the WD cycles, reaching the maximum with 15% lime. The highest increase of 49% (4.6 to 6.8 MPa) was recorded between the strength before and after the WD cycles.
- The compressive strength further increases from 4.4 to 8.4 MPa for the CEBs stabilized with 20:0 to 12:8% lime:ash, reaching the maximum with 16:4% lime:ash, before the WD cycles. It further increases from 6.2 to 9.5 MPa after the WD cycles. It recorded the highest increase of 40% (4.4 to 6.2 MPa) with 20:0% lime:ash, from before to after the WD cycles, highlighting the resistance to the WD cycles of CEBs stabilized with lime-rich binders.
- On the contrary, the compressive strength decreases for all content of cement after the WD cycles. The highest decrease reached 55% (12.5 to 5.5 MPa) for the CEBs stabilized with 4% cement, confirming the lack of the resistance to the WD cycles of the CEBs stabilized with cement.
- However, the compressive strength increases by up to 58% (4.2 to 6.6 MPa) for the CEBs stabilized 8% cement after exposure to the temperature of up to 600°C. This suggests that the CEBs can withstand the relatively high temperature without risking the loss of their mechanical performance.

This study shows that the mechanical performances of CEBs are influenced by the exposure to WD drying cycles and high temperature (up to 600°C) and the type of stabilizer. It is recommended that more studies assess other engineering and durability performances of CEBs and stabilized with different types of chemical binders, and specifically at the temperature higher than 600°C. This would allow to confirm whether the increase of the compressive strength is maintained beyond 600°C.

ACKNOWLEDGEMENTS

"Académie de la Recherche et de l'Enseignement Supérieur" of the "Fédération Wallonie – Bruxelles (Belgium) – Commission de la Coopération au Développement" (ARES-CCD) provided the financial support as part of an international research and development project "Amélioration de la qualité de l'habitat en terre crue au Burkina Faso (Improving the quality of earth-based habitats in Burkina Faso) PRD2016-2021". Burkina Industrial Gas (BIG) is acknowledged for generously providing the lime residue.

REFERENCES

[1] Zoungrana, O., Bologo/Traore, M., Hema, C., Nshimiyimana, P., Pirotte, G. & Messan, A., Sustainable habitat in Burkina Faso: Social trajectories, logics and motivations for the use of compressed earth blocks for housing construction in Ouagadougou. *WIT Transactions on The Built Environment*, vol. 195, WIT Press: Southampton and Boston, pp. 165–172, 2020. DOI: 10.2495/ARC200131.

[2] Zoungrana, O., Bologo/Traore, M., Messan, A., Nshimiyimana, P. & Pirotte, G., The paradox around the social representations of compressed earth block building material in Burkina Faso: The material for the poor or the luxury material? *Open J. Soc. Sci.*, **9**, pp. 50–65, 2021. DOI: 10.4236/jss.2021.91004.

[3] Sore, S.O., Messan, A., Prud'homme, E., Escadeillas, G. & Tsobnang, F., Stabilization of compressed earth blocks (CEBs) by geopolymer binder based on local materials from Burkina Faso. *Constr. Build. Mater.*, **165**, pp. 333–345, 2018. DOI: 10.1016/j.conbuildmat.2018.01.051.

[4] Tarmangue, D., Sore, S.O., Nshimiyimana, P., Messan, A. & Courard, L., Comparative study of the reactivity of clay earth materials for the production of compressed earth blocks in ambient conditions: Effect on their physico-mechanical performances. *J. Miner. Mater. Characteriization Eng.*, **10**, pp. 40–56, 2021.

[5] Nshimiyimana, P., Hema, C., Zoungrana, O., Messan, A. & Courard, L., Thermophysical and mechanical properties of compressed earth blocks containing fibres: By-product of okra plant and polymer waste. *WIT Transactions on The Built Environment*, vol. 195, WIT Press: Southampton and Boston, pp. 149–161, 2020. DOI: 10.2495/ARC200121.

[6] Hema, C., Soro, D., Nshimiyimana, P., Lawane, A., Messan, A. & Van Moeseke, G., Improving the thermal comfort in hot region through the design of walls made of compressed earth blocks: An experimental investigation. *J. Build. Eng.*, **38**, 102148, 2021.

[7] Hema, C., Messan, A., Lawane, A. & Van Moeseke, G., Impact of the design of walls made of compressed earth blocks on the thermal comfort of housing in hot climate. *Buildings*, **10**, p. 157, 2020. DOI: 10.3390/buildings10090157.

[8] Turco, C., Paula Junior, A.C., Teixeira, E.R. & Mateus, R., Optimisation of compressed earth blocks (CEBs) using natural origin materials: A systematic literature compilation review. *Constr. Build. Mater.*, **309**, 125140, 2021. DOI: 10.1016/j.conbuildmat.2021.125140.

[9] Nshimiyimana, P., Sore, S.O., Hema, C., Zoungrana, O., Messan, A. & Courard, L., A discussion of "Optimisation of compressed earth blocks (CEBs) using natural origin materials: A systematic literature review". *Constr. Build. Mater.*, **309**, 2022. DOI: 10.1016/j.conbuildmat.2021.125140.

[10] Nshimiyimana, P., Moussa, S.H., Messan, A. & Courard, L., Effect of production and curing conditions on the performance of stabilized compressed earth blocks: Kaolinite vs quartz-rich earthen material, *MRS Adv.*, **5**, pp. 1277–1283, 2020. DOI: 10.1557/adv.2020.155.

[11] Beckett, C. & Kazamias, K., A. Law, Inv|ESTIGATIONS into the high temperature behaviour of unstabilised rammed earth. *Sustain. Constr. Mater. Technol.*, **3**, 2019. DOI: 10.18552/2019/idscmt5161.

[12] Abdallah, R., Perlot, C., Carré, H., La Borderie, C. & El Ghoche, H., Fire behavior of raw earth bricks: Influence of water content and cement stabilization. **1**, pp. 792–800, 2022.

[13] Bakam, V.A., Mbishida, M.A., Danjuma, T., Zingfat, M.J., Hamidu, L.A.J. & Pyendang, Z.S., Effect of firing temperature on abrasive and compressive strengths of an interlocking compressed stabilized earth block (CSEB). *Int. J. Recent. Eng. Sci.*, **7**, pp. 44–46, 2020. DOI: 10.14445/23497157/ijres-v7i3p109.

[14] Hakimi, A., Ouissi, H., El Kortbi, M. & Yamani, N., Un test d'humidification-séchage pour les blocs de terre comprimée et stabilisée au ciment. *Mater. Struct.*, **31**, pp. 20–26, 1998.

[15] Yogananth, Y., Thanushan, K., Sangeeth, P., Coonghe, J.G. & Sathiparan, N., Comparison of strength and durability properties between earth-cement blocks and cement–sand blocks. *Innov. Infrastruct. Solut.*, **4**, pp. 1–9, 2019. DOI: 10.1007/s41062-019-0238-8.

[16] Nshimiyimana, P., Fagel, N., Messan, A., Wetshondo, D.O. & Courard, L., Physico-chemical and mineralogical characterization of clay materials suitable for production of stabilized compressed earth blocks. *Constr. Build. Mater.*, **241**, 118097, 2020. DOI: 10.1016/j.conbuildmat.2020.118097.

[17] Nshimiyimana, P., Miraucourt, D., Messan, A. & Courard, L., Calcium carbide residue and rice husk ash for improving the compressive strength of compressed earth blocks. *MRS Adv.*, **3**, pp. 2009–2014, 2018. DOI: 10.1557/adv.2018.147.

[18] Nshimiyimana, P., Messan, A. & Courard, L., Physico-mechanical and hygro-thermal properties of compressed earth blocks stabilized with industrial and agro by-product binders. *Materials (Basel)*, **13**, p. 3769, 2020. DOI: 10.3390/ma13173769.

[19] PR XP P13-901, Blocs de terre comprimée pour murs et cloisons: Définitions – Spécifications – Méthodes d'essai – Conditions de réception, 2017.

[20] Nshimiyimana, P., Messan, A., Zhao, Z. & Courard, L., Chemico-microstructural changes in earthen building materials containing calcium carbide residue and rice husk ash. *Constr. Build. Mater.*, **216**, pp. 622–631, 2019. DOI: 10.1016/j.conbuildmat.2019.05.037.

[21] Thiruvadi, M. & Reddy, A.N., Investigation on metakaolin and silicafume incorporated concrete under elevated temperature. *Conference Proceedings of the 2nd Int Conf Adv Eng Appl Sci Manag.*, 2018.

[22] Fernandes, B., Carré, H., Mindeguia, J.-C., Perlot, C. & La Border, C., Fire spalling sensitivity of concrete made with recycled concrete aggregates (RCA). *Acta Polytech CTU Proc.*, **33**, pp. 168–174, 2022. DOI: 10.14311/APP.2022.33.0168.

INTEGRATIVE REINFORCEMENT OF NATURAL FIBER: FABRICATION OF PEANUT SHELL TILES IN BUILDING CONSTRUCTION

DEENA EL-MAHDY*
The British University in Egypt, Egypt

ABSTRACT

Alternative recycling materials are one of the most promising strategies in the construction sector for reducing global warming, high energy demand, and carbon emission release. Several studies oriented their research toward encouraging the concept of recycling, reusing, and reducing agriculture plant waste to be integrated into material construction. With a focus on "peanut shell waste", the paper aims to highlight the possibility of generating lightweight and sturdy units/tiles in building construction. The method first started with evaluating the energy parameters effect of peanut shells in building construction, followed by the experimental test. This test was done by examining a mix of peanut shells with sand and corn starch to generate tiles/brick units. Different form-generation were explored to fabricate the casting mould of the unit. With the use of bio-resin, the mix was cast in the mould and then assembled after drying. The resulting tile's prototype showed the possibility of using them as 3D units that can be assembled to generate a self-structured element, or as tiles to be used indoors or outdoors. Different architecture applications can be generated from these tiles/units as an alternative to industrial building materials which can decrease the environmental impact of natural waste storage. In addition, these tiles will enhance the thermal performance of the space if it is used as insulation in the walls, which in parallel will reduce carbon emission. Further research is to generate 3D printing filament from peanut shell waste as bio-based material. Another study can assess the difference between peanut shells mixed with clay bricks with traditional fired bricks from a mechanical behaviour and thermal performance point of view for humidity and temperature monitoring.

Keywords: peanut shell, recycle, casting, form-finding, digital fabrication.

1 INTRODUCTION

The massive increase in population growth increases the demand for new buildings and more agricultural industries. The increase of both industries resulted in generating a huge amount of abundant materials and agricultural waste that affect the environment if burnt or left to dispose of the land. However, both processes release huge amounts of greenhouse gases into the environment. Therefore, it is imperative to find ways to reuse and recycle this crop waste. Industrial materials in construction sectors are responsible for the climate change crisis resulting in high energy demand and carbon emission release which affect human thermal comfort. Several studies have shown the possible application of agricultural wastes in building materials enhancing their mechanical properties and reducing the cost of materials production, pollution, and gas emissions [1]. However, a lack of studies on durability and thermal properties is observed. Providing alternative waste management through reusing it can help in reducing environmental impacts worldwide [2].

The research question highlight how crop waste can be recycled into building materials, thereby minimizing waste and improving sustainability within the construction industry. As a way of exploring the possibility of using peanut-shell waste as a raw material without burning, the paper studied the generation of lightweight units/tiles in building construction. The method first started with reviewing previous studies that highlighted the peanut shells

* *ORCID: https://orcid.org/0000-0001-8755-8733*

WIT Transactions on The Built Environment, Vol 210, © 2022 WIT Press
www.witpress.com, ISSN 1743-3509 (on-line)
doi:10.2495/ARC220131

properties, energy effect and carbon emission released from, followed by an experimental test. Two tests were done; first to examine the behaviour of peanut shell when it mixed with sand and corn starch, second, using bio-resin without any heat or pressure. Form-generation test was explored to fabricate a mould to cast the mixture in then assemble the units after drying. The research develops a vision of global challenge translated in an architectural application in form of tiles extracted from agricultural waste as an alternative in today's construction material industry. The prototype obtained renewability, sustainability, and innovation character.

2 RELATED WORK

2.1 Groundnut/peanut shell production worldwide

The continuous global population growth resulted in the increase demand in agriculture and construction that provide both food and shelters, which in parallel increased more waste and pollution to the environment [3]. To get rid of the agriculture plant waste, many countries used to burn them which harms the environment by polluting the air through the gas emission produced from this process. Groundnut shells waste for instance are burnt or buried resulting in environmental pollution [4]. In 1997, Egypt witnessed one of the biggest black cloud phenomena that cover the Delta and Cairo resulting from burning the of rice straw. After the great toxic emission resulting from this phenomenon, the ministry of agriculture with the ministry of environment banned burning the rice straw in the open-air and put a protocol that encourage reusing and recycling the rice straw for many organic products and fuels which provide more jobs. Agricultural waste provides more than 33% of total biofuels usage in the world [5]. This waste includes fibres and bio-composite that can be used as an alternative for architectural construction materials [6]. Generally, with the shortage in the resources resulting from the climate change problems, searching for clean ways to recycle the plant waste has been received increasing attention in several fields. Regardless of the high energy consumption resulting from the building sectors, especially in cement production, which is the main binder used in concrete, mortar, and fired bricks, had led to search for an alternative sustainable construction materials for reducing/replacing cement [7]. This can lead to more sustainable industry, not only as a way to reduce the high cost but to reduce the high energy demand and the CO_2 emission released from cement production which reaches 5% [8].

As for the research focus, peanut, which is also known as groundnut, is used mainly in food production that contains many nutritional minerals and is imported in three different forms; in-shell, shelled, and peanut oil [9]. Based on Atlasbig, the largest countries of peanut production are China in the first place with a total of 17.5 million tonnes per year, followed by India, and Nigeria with 6.7 and 4.5 million tonnes per year respectively [10]. Together they are responsible for more than 50% of the world's total peanut production globally as shown in Fig. 1. Egypt ranked at 22 with a total of 231,223 tonnes per year where 30%–35% of the total production is exported globally. While the largest countries for importing peanuts are Italy, Germany, Greece, and the Netherlands. With all these amount of peanut production, tons of peanut shells are discarded to be a huge waste, where different solutions need to be done to reuse these waste [11].

As a way of reusing raw crop waste, particleboard is one of the most famous applications that generates panel under pressure and heat from wood or lignocellulosic fibres combined with resins and bonded together. This process is considered economic and low-cost where the furniture industry benefits from these boards. Other applications of these particleboards

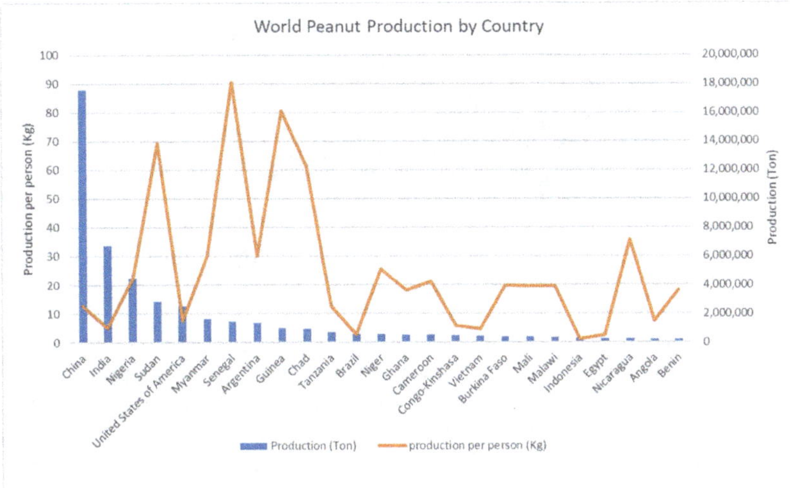

Figure 1: The world peanut production.

can be seen in walls, partitions, tiles, roof top, etc. [12]. Several types of resins are used in manufacturing these boards for instance, phenol formaldehyde (PF) which is known with its high strength, thermal and moisture resistance, excellent temperature stability, less toxic, and high compaction of the fibres [12].

2.2 Architecture applications of peanut shell and their mechanical properties

Many studies have been focusing on integrating agriculture crop waste among the natural fibres into traditional building materials and structural composites because their higher strength and stiffness with the availability considering a waste [13]–[15]. Natural fibres advantages can be seen in their lower pollutant and greenhouse emissions, lower thermal expansion, lightweight, less cost, less energy demand and more recovery, eco-friendly, and biodegradable components at the end of the building life-cycle [16], [17]. These wastes are used as an alternative to reduce the amount of the cement and clay used in concrete and bricks manufacturing reaching more strength and lightness. Crop wastes capture CO_2 during growth, and their harvesting does not use large amounts of fuel energy [15]. Their growth absorb CO_2 with less amount of energy needed during harvesting, and an average of 5–10 times less non-renewable energy than glass fibres [16], [18] and 30–70 times less than carbon fibres [19] during production.

Both biomaterials extracted from plant waste and carbon fibres became the leading materials used especially in structure. At Stuttgart University at ITKE/ICD, a team led by Achim Menges and Jan Knippers and their team constructed two pavilions at Stuttgart University based on biomimicry principles. They used woven-resin-impregnated glass and carbon fibre. These materials allow generating lightweight composite that reached a large-span structure to reach the aim of reducing the formwork as shown in Fig. 2 [20]. The woven technique by the robot arm allowed seamless and joint less interaction for the whole structure. Dahy and her team [21] at BioMat/ITKE based in Stuttgart university generated the BioMat pavilion, where they used modular bio composite sandwich panels combined with the structural beams. Also, she generated the Polyclad Acoustic Panel out of bio composite and thermoplastic as a way of recycling agricultural waste.

Figure 2: Carbon-fibre pavilion, ITKE/ICD, Stuttgart university, 2017.

Various researchers tested different applications of using peanut shell, include its uses as building material insulation mixed with plaster evaluating its thermal properties [7], [14]. achieved good compression strength through using groundnut shell ash as a partial replacement of cement in sand Crete blocks with an average of 10% to 20% percentage with compression strength average of 3.8 N/mm2. Quaranta et al. [22] used peanut shell for ceramic manufacture with percentage of 5%, 10% and 15%, where it shows the more percentage of the biomass, the higher energy needed to get good strength and high compressive strength. Adeosun et al. [23] used groundnut shells in nanofiber preparation which were used in packing and tissue engineering applications [12] generated medium-density panel from peanut shell after testing different fibres percentage of 50%, 60%, 70% and 80% through compaction. Tensile strength showed higher value of 11.4 MPa in the ratio of 50%. For achieving light weight concrete. Sada et al. [24] used efficiently groundnut shell to reduce the amount of the cement used in concrete production. Also, Mahmoud et al. [7] used the shells for replacement of the fine aggregates in the concrete with reaching percentage of 75%. Escalera et al. [25] produced lightweight porous fired bricks using nut shells where a decreasing shrinkage were observed from 11.60% to 6.80% with an increase in the thermal conductivity from 0.20 W/mK to 0.76 W/mK [2].

2.3 Energy and recyclability

Perea-Moreno et al. [26] reports in their study that the energy properties of the peanut shells contribute for environmental conservation because of their lower emission comparing to biomass, with the low content of sulphur 0.54%, nitrogen 0.50% and the chlorine 0.07%. This make the peanut shell ideal for biofuel that would decrease the CO_2 emission to be used in production either for heating or cooling. The same study recorded very low value of moisture content in the peanut shell with a value of 5.79%, which make an advantages that no drying treatment is needed using any energy purposes [26]. A study done by Perea-Moreno et al. shows that biofuel is the highest production extracted from the peanut shells globally, where China is the greatest to extract the energy from it follow by India, Nigeria, the US, and Sudan with average of 25,579.75, 11,440.42, 5253.69, 3637.60 and 2891.79 MWh respectively.

3 METHODOLOGY

3.1 Materials and tools

The material used in with the peanut shells are sand, corn-starch, and bio resin. Based on a study done by Kulshreshtha et al. [27], when corn starch is mixed with water and sand, they forms a non-Newtonian fluid state where the starch granules in cold water. When this mixture is heated in oven or microwave for 3 min, it generates a hardened mix called CoRncrete [27]. This composition has a compressive strength similar to the fired clay bricks. In addition to this mix, this research integrated the peanut shells as recycled materials that act as fibres by grinding them into two different sizes: coarse and fine grains of 6–8 mm and 2–4 mm respectively as shown in Fig. 3. The sand and corn starch particles varied between 0.5 and 1 mm, which act as binder to bond the peanut shells with the mix in form of 2 cubes.

Figure 3: Different size of grinded peanut shells.

Bio-resin was added to the mixture instead of corn starch and water where no heating, which acts as a binder and gives more strength to the tiles. The mix was cast in a mould and assembled after drying in room temperature. Scale, grinder machine, and microwave were used. The resin used in the research is a bio-based epoxy that has a density of 1.09 g/cm3, tensile strength of 67.5 MPa, energy intensity of 49 MJ/kg, and global warming potential of 4.08 kg CO_2-eg/kg [15].

3.2 Method

In order to generate peanut shell tiles, the literature review aimed to evaluate the energy effect resulted from the peanut shells in building construction, followed by the experimentation test. This test is divided into three main phases: (1) material characterization, (2) form-finding, (3) fabrication of mould, and (4) the casting and assembly process as shown in Fig. 4. The peanut shells were collected and grained into coarse and fine particles and sieved as powder. Two different of binders were tested; cornstarch that need heating, and bio-resin the does not need any heating process.

3.2.1 Material characterisation test

To study the microstructure of the powder shells, SEM images were taken using Olympus microscope model no. BX41 showing both the outer skin protection layer and the fibres inside. Two cubic moulds were made to test the homogeneous of the mixture with a ratio of sand, cornstarch, and shell (1:2:2) for both fine and coarse shells. The samples were heated in a microwave with temperature of 200°C to bond the cornstarch which acts as a binder. Bio resin was replaced by the cornstarch to remove the heating process with a ratio to the sand and shells of (0.5:1:2). Compression test was done for the samples.

3.2.2 Form generation

The unit's form was inspired by the outer shell of the peanut morphology. The form consists of two elliptical shapes from both sides. This elliptical shape was developed to give

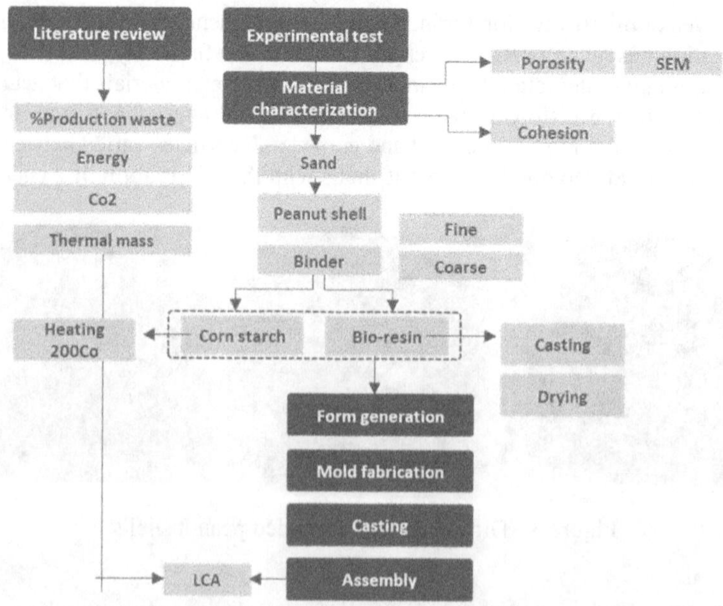

Figure 4: The method of the research.

alternative and multiple forms as shown in Fig. 5(a) to reach a unit that can be connected and attached to another either overlapped or side by side. The arrangement of the units were in two directions, where the small units are perpendicular and the bigger units are diagonal to assemble similar form of hexagon.

Two squared sizes were designed considering the connection between the units 12 cm and 8 sm. The bigger units are assembled diagonally, while smaller units are assembled perpendicular which acted as connectors as shown in Fig. 5(b).

(a) (b)

Figure 5: (a) The form finding of the unit; and (b) The arrangement and assembly of both units.

3.2.3 Mould fabrication and casting process

Two different sizes of mould were cut using acrylic sheet with thickness of 5 mm using laser cutter machine to cast the mixture in as shown in Fig. 6. Two mixtures were prepared one for the fine grains and the other for the coarse grains to generate different porosity which will affect the lightness of the tiles. The peanut shells are mixed with sand and resin as a binder.

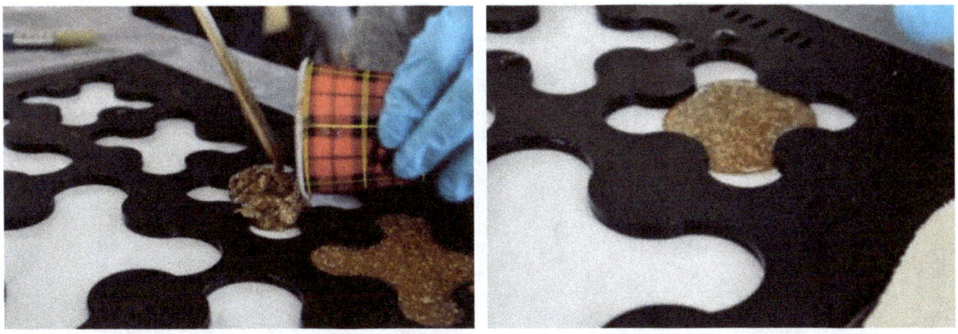

Figure 6: Two units' sizes during casting process in the mould.

The mixture has filled the units after reaching a homogeneous state. The casted units are kept drying within few hours in room temperature 24°C.

4 RESULTS AND DISCUSSION

4.1 Material behaviour

SEM images in Fig. 7 shows the microstructure of the peanut shell for both pattern and porosity. While the fine particles have less spaces in between which make it denser. This concludes that the small particles of both sand and corn-starch can fill the space faster, which will need more materials in case the bigger particles as shown in Fig. 8. For the mechanical behaviour, the cubic samples for both fine and coarse grains recorded compressive strength of 28 to 22 MPa, respectively. It was noticed that the cohesion of the coarse sample was too lose.

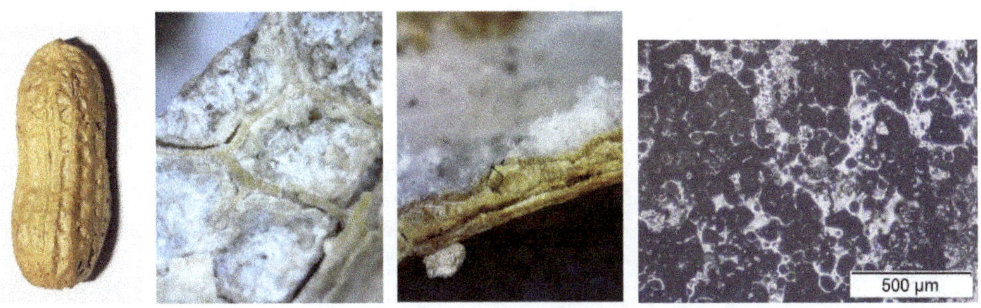

Figure 7: The external peanut shell and SEM images shows pattern of the shell and the microstructure of the peanut shell thickness.

For the samples with bio-resin, the mixture does not need any heating or pressuring process, and the solidification hardened in less than two hours. Unlike the cubic samples, it was observed that the compressive strength of the fine grains to the coarse grains was 50 MPa and 62 MPa. Yet in case of using the powder as 3D printing material or turning it into filament, heat will be needed to melt the binder which will increase the compression strength when resin released to temperature between 500 and 1000°C.

Figure 8: Two samples after heating using corn starch with fine and coarse shells.

4.2 Porosity of the tessellated unit

It is easy to control the openings' size in the block to give more variety depending on the function beneath. Fig. 9(a) shows the variation in the opening's size and how the unit be a 3D which will use differently. These openings in the case of wall partitions can be changed with different sizes depending on the function. The porosity will result more lightweight tiles. Connections between the tiles leave some voids in a form of square as shown in Fig. 9(b) which can be used as a ventilation system or light entry.

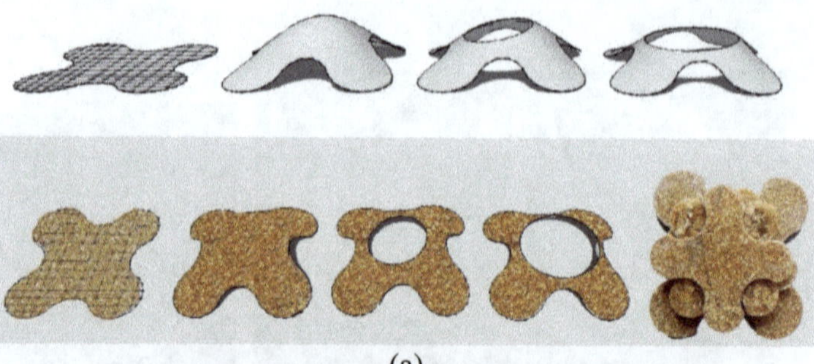

(a)

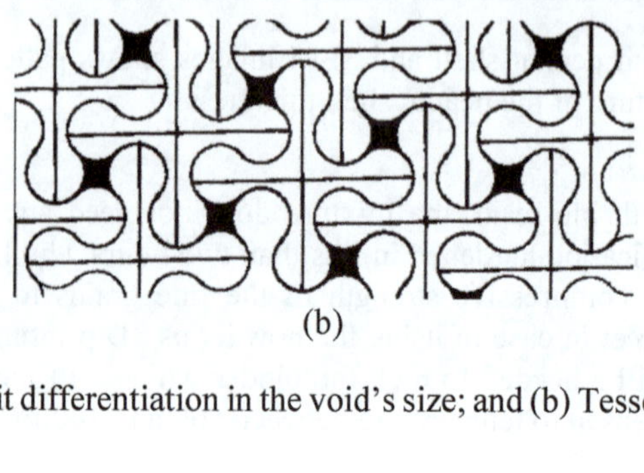

(b)

Figure 9: (a) The unit differentiation in the void's size; and (b) Tessellation of the proposed unit.

4.3 Connection and fabrication assembly

The unit can be connected in several ways giving different alternatives. The join can be through sticking the units together overlapped, or through steel rebar where the units can form the final panel. The overlapping forms when using in walls give the ability to reduce temperature inside the space, which act as an insulator.

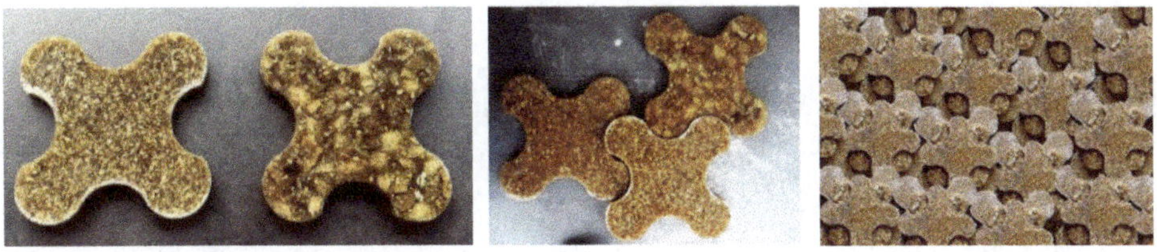

Figure 10: The generated units; fine and coarse shells, the connection, and the panel assembly.

4.4 Architecture application and life cycle assessment

Many architectural applications can integrate the peanut shells either in interiors as cladding and finishes, wall system or in outdoors such as in tiles, panels, and boards that can be reused in furniture design. For building materials, peanut shells can be used as insulation filler in air cavity to reduce the thermal resistance of the wall causing temperature heat reduction which is left for another research. Another application can be done as fibres in fired brick, sun dried bricks, mortar, or in concrete production as a replacement of the cement production. Additionaly, cellioluse extraction can be produced out for other bio-composite prodcution. For life cycle assessment, the bio-resin can be melter and reused again, yet, some carbon emession will release.

5 CONCLUSION

The paper presented the possibility of using agricultural waste as a base in building material. The resulted units could act as a friendly material that can be replaced with the non-efficient material that harms the environment. The literature presented some results of previous research how the shells out of crops assess in reducing the waste in the environment instead of burning them. After different form findings, the paper showed the possible alternatives that can be generated from one form. Porosity of the particles in the coarse or the fine state showed differentiation in the weight that respond back to the mixture cohesion. The arrangement and the connection of the units through using steel rebars can be done in further research testing the structural stability of panel application. As the paper presented an application of using the waste without any heating or pressure, further research could be implemented to turn this waste into filament material for 3D printer, which later can be a door for multiple applications through different additive manufacturing methods. The units can be applied in architecture as an alternative construction building material which could reduce the total carbon emission and energy needed for the building. Life cycle assessment will be done in another part as a continue of this paper.

ACKNOWLEDGEMENTS

The author would like to thank Yask Kulshreshtha who helped with the cornstarch assessment and the team who shared in part of the experimentation test: Ahmed Khairy, Annina Moser, Balaj Fazal, and Daniel Turk. This project was part of the Engineering for Development (E4D) "Sand: an (in)finite resource?" Summer School funded by the Sawiris Foundation for Social Development in Egypt, and hosted in TU Delft with collaboration with ETH Zurich.

REFERENCES

[1] Ortiz, O., Castells, F. & Sonnemann, G., Sustainability in the construction industry: A review of recent developments based on LCA. *Constr. Build. Mater.*, **23**(1), pp. 28–39, 2009. DOI: 10.1016/j.conbuildmat.2007.11.012.

[2] Jannat, N., Al-Mufti, R.L., Hussien, A., Abdullah, B. & Cotgrave, A., Utilisation of nut shell wastes in brick, mortar and concrete: A review. *Constr. Build. Mater.*, **293**, pp. 1–17, 2021.

[3] Ongpeng, J.M.C., Inciong, E., Sendo, V., Soliman, C. & Siggaoat, A., Using waste in producing bio-composite mycelium bricks. *Appl. Sci.*, **10**(15), pp. 1–13, 2020. DOI: 10.3390/APP10155303.

[4] Duc, P., Dharanipriya, P., Velmurgan, B. & Shanmugavadivu, M., Groundnut shell: A beneficial bio-waste. *Biocatal. Agric. Biotechnol.*, **20**, pp. 1–407, 2009. DOI: 10.1201/9781420077070.

[5] Yevich, R. & Logan, J.A., An assessment of biofuel use and burning of agricultural waste in the developing world. *Global Biogeochem. Cycles*, **17**(4), pp. 1–42, 2003. DOI: 10.1029/2002gb001952.

[6] Ghazvinian, A., Farrokhsiar, P., Vieira, F., Pecchia, J. & Gursoy, B., Mycelium-based bio-composites for architecture: Assessing the effects of cultivation factors on compressive strength. *eCAADe 37 – Matter – Materials Studies and Innovation*, **2**, pp. 505–514, 2020. DOI: 10.5151/proceedings-ecaadesigradi2019_465.

[7] Mahmoud, H., Belel, Z.A. & Nwakaire, C., Groundnut shell ash as partial replacement of cement in sandconcrete blocks production. *Int. J. Dev. Sustain.*, **1**(3), pp. 1026–1032, 2012. DOI: 10.5958/2349-2988.2017.00056.0.

[8] Worrell, E., Price, L., Martin, N., Hendriks, C. & Meida, L.O., Carbon dioxide emissions from the global cement industry. *Annu. Rev. Energy Environ.*, **26**, pp. 303–329, 2001. DOI: 10.1146/annurev.energy.26.1.303.

[9] Prusak, A.M., Schlegel-Zawadzka, M., Boulay, A. & Rowe, G., Characteristics of the peanut chain in Europe: implications for peanut allergy. *Acta Sci. Pol. Technol. Aliment.*, **13**(3), pp. 321–333, 2014. DOI: 10.17306/J.AFS.2014.3.10.

[10] AtlasBig, Countries peanut production. 2021. https://www.atlasbig.com/en-us/world-energy-statistics. Accessed on: 2 May 2022.

[11] Bai, S., Wang, T., Tian, Z., Cao, K. & Li, J., Facile preparation of porous biomass charcoal from peanut shell as adsorbent. *Nature Scientific Reports*, **10**, 15845, 2020. DOI: 10.1038/s41598-020-72721-0.

[12] Batalla, L., Nuñez, A.J. & Marcovich, N.E., Particleboards from peanut-shell flour. *J. Appl. Polym. Sci.*, **97**(3), pp. 916–923, 2005. DOI: 10.1002/app.21847.

[13] Miao, M. & Finn, N., Conversion of natural fibres into structural composites. *J. Text. Eng.*, **54**(6), pp. 165–177, 2008. DOI: 10.4188/jte.54.165.

[14] Lamrani, M., Laaroussi, N., Khabbazi, A., Khalfaoui, M., Garoum, M. & Feiz, A., Experimental study of thermal properties of a new ecological building material based on peanut shells and plaster. *Case Stud. Constr. Mater.*, **7**, pp. 294–304, 2017. DOI: 10.1016/j.cscm.2017.09.006.
[15] Gil Pérez, M., Guo, Y. & Knippers, J., Integrative material and structural design methods for natural fibres filament-wound composite structures: The livmats pavilion. *Mater. Des.*, **217**, 2022. DOI: 10.2139/ssrn.4010392.
[16] Joshi, S.V., Drzal, L.T., Mohanty, A.K. & Arora, S., Are natural fiber composites environmentally superior to glass fiber reinforced composites? *Compos. Part A Appl. Sci. Manuf.*, **35**(3), pp. 371–376, 2004. DOI: 10.1016/j.compositesa.2003.09.016.
[17] Xu, X., Jayaraman, K., Morin, C. & Pecqueux, N., Life cycle assessment of wood-fibre-reinforced polypropylene composites. *J. Mater. Process. Technol.*, **198**(1–3), pp. 168–177, 2008. DOI: 10.1016/j.jmatprotec.2007.06.087.
[18] Song, Y.S., Youn, J.R. & Gutowski, T.G., Life cycle energy analysis of fiber-reinforced composites. *Compos. Part A Appl. Sci. Manuf.*, **40**(8), pp. 1257–1265, 2009. DOI: 10.1016/j.compositesa.2009.05.020.
[19] Das, S., Life cycle assessment of carbon fiber-reinforced polymer composites. *Int. J. Life Cycle Assess.*, **16**(3), pp. 268–282, 2011. DOI: 10.1007/s11367-011-0264-z.
[20] Reichert, S., Schwinn, T., La Magna, R., Waimer, F., Knippers, J. & Menges, A., Fibrous structures: An integrative approach to design computation, simulation and fabrication for lightweight, glass and carbon fibre composite structures in architecture based on biomimetic design principles. *CAD Comput. Aided Des.*, **52**, pp. 27–39, 2014. DOI: 10.1016/j.cad.2014.02.005.
[21] Dahy, H., Natural fibre-reinforced polymer composites (NFRP) fabricated from lignocellulosic fibres for future sustainable architectural applications, case studies: Segmented-shell construction, acoustic panels, and furniture. *Sensors*, **19**(3), 2019. DOI: 10.3390/s19030738.
[22] Quaranta, N., Caligaris, M., Pelozo, G., Césari, A. & Cristóbal, A., Use of wastes from the peanut industry in the manufacture of building materials. *Int. J. Sustain. Dev. Plan.*, **13**(4), pp. 662–670, 2018. DOI: 10.2495/SDP-V13-N4-662-670.
[23] Adeosun, S., Taiwo, O., Akpan, E., Gbenebor, O., Gbagba, S. & Olaleye, S., Mechanical characteristics of groundnut shell particle reinforced polylactide nano fibre. *Rev. Mater.*, **21**(2), pp. 482–491, 2016. DOI: 10.1590/S1517-707620160002.0045.
[24] Sada, B.H., Amartey, Y.D. & Bako, S., An investigation into the use of groundnut shell as fine aggregate replacement. *Niger. J. Technol.*, **32**(1), pp. 54–60, 2013. DOI: 10.4314/NJT.V32I1.
[25] Escalera, E., Garcia, G., Terán, R., Tegman, R., Antti, M.L. & Odén, M., The production of porous brick material from diatomaceous earth and Brazil nut shell ash. *Constr. Build. Mater.*, **98**, pp. 257–264, 2015. DOI: 10.1016/j.conbuildmat.2015.08.003.
[26] Perea-Moreno, M.-A., Manzano-Agugliaro, F., Hernandez-Escobedo, Q. & Perea-Moreno, A.-J., Peanut shell for energy: Properties and its potential to respect the environment. *Sustainability*, **10**, pp. 1–15, 2018. DOI: 10.3390/su10093254.
[27] Kulshreshtha, Y., Schlangen, E., Jonkers, H.M., Vardon, P.J. & van Paassen, L.A., CoRncrete: A corn starch based building material. *Constr. Build. Mater.*, **154**, pp. 411–423, 2017. DOI: 10.1016/j.conbuildmat.2017.07.184.

GRAPHENE IN ARCHITECTURE

MOEMEN MOHAMED EZZ & MOHAMED ABD EL AL IBRAHIM ZEYAD TAREK EL SAYA
Department of Architecture Engineering, Alexandria University, Egypt

ABSTRACT
Graphene comprises of a single layer of sp2 hybridized carbon iotas shaping a two dimensional (2D) hexagonal cross section. Graphene incorporated-polymer nanocomposites appear more noteworthy mechanical, thermal, and electrical properties compared to flawless polymer which can discover applications within the fields of car, aviation, gadgets and green vitality. Graphene, with which Andréy Gueim and Konstantín Novosiólov were granted the Nobel Prize of Material Science in 2010, is beginning to see its applications within the genuine world. Graphenano, headquartered in Yecla (Spain), outlined a strategy to make this fabric on a mechanical scale. The tall footing and tearing resistance of graphene make it the "ideal additive" for cement and concrete. The item aims to progress all the benefits that "affect the toughness of concretes and compromise their great execution as time passes by". It is significant to discover more ecologic opportunities for development for the reduction of carbon outflows around the world.
Keywords: energy efficiency, sustainability, nanotechnology, graphene, graphene coating, solar photovoltaic, thermal conductivity.

1 INTRODUCTION

Despite it only being published in 2004, the seminal work "Electric Field Impact in Molecularly Lean Carbon Films" has been cited more than 15,000 times and is at number 65 within the best 100 list, The *Nature* article is given a comparison with Mount Everest where these 100 papers are compared to a 1 cm thick tip at the best of the number of papers distributed over a long time. Manchester analysts disconnected graphene at The College of Manchester in 2004, and two of them, Sir Andre Geim and Sir Kostya Novoselov, were awarded the Nobel Prize for Material Science for illustrating graphene's extraordinary properties set out within the paper [1]. Graphene may be a two-dimensional single-atom thick film of carbon iotas orchestrated in a honeycomb gem. It could be an ideal case of a two-dimensional electron framework for a physicist, an exquisite shape of a two-dimensional natural macromolecule comprising of benzene rings for a chemist and a fabric with monstrous conceivable outcomes for an build due to its fabulous electrical, magnate and thermal properties [2].

2 REVELATION OF GRAPHENE NANOMATERIALS

The word graphene is begun from the Greek word "Graphene" which means to type in. Previous inquire about, cantering on arrangement and essential properties of nanocarbons (e.g., epitaxial graphene movies, nanoribbons and nano patches), gives an essential crucial information. Graphene's inquire about has been effectively capitalised after its to begin with transformation by Novoselov et al. in 2004. Graphene's being an interesting or advancement form of 2-dimensional (2D) to 3D topology, have potential properties and Fig. 1 shows one of the biggest institutes of graphene. These 3D materials can be functionalised by the substituting the carbon iotas with heteroatoms or whole utilitarian bunches [3].

WIT Transactions on The Built Environment, Vol 210, © 2022 WIT Press
www.witpress.com, ISSN 1743-3509 (on-line)
doi:10.2495/ARC220141

Figure 1: Henry Royce Institute in UK. *(Source: https://www.graphene.manchester.ac.uk/about/henry-royce-institute/.)*

2.1 History of graphene

The 19th century saw numerous wondrous disclosures, basic disclosures, unused materials, electromagnetic marvels. English chemist Benjamin Collins Brodie recognized the exceedingly layered nature of thermally diminished graphite oxide as early as 1859 having detailed the nuclear weight of graphite within the Philosophical Exchanges of the Illustrious Society of London that year. Powder diffraction was utilized to fathom the structure of graphite in 1916 and Kohlschütter and Haenni portrayed what they called graphite oxide "paper" in 1918. In attempting to create a hypothesis for the electronic properties of bulk graphite, P R Wallace supported his work with a hypothesis of graphene in 1947 [3].

2.2 Types of graphene

2.2.1 Hydrogenated graphene

A fully hydrogenated derivative of graphene, of which the chemical composition is CH, is predicted based on first-principles total energy calculations and named graphene. It consists of sp3 C–C bonds, as opposed to graphene's sp2 bonds, and consequently a carbon atom layer is puckered. Graphene can have two conformations, as schematically shown on their top-views in Fig. 2 a chair-like conformer with the hydrogen atoms alternating on both sides of the carbon atom layer and a boat-like conformer with the hydrogen atoms alternating in pairs The space group, bond distances and binding energy predicted6 are shown for two In the chair-type conformer, the calculated C–C bond length of 0.152 nm is similar to that in diamond (sp3 bonding) and is much greater than that in graphene (0.142 nm characteristic for sp2 bonding [4].

2.2.2 Fluorinated graphene (fluorographene)

Although promising performances as cathode in primary lithium battery, only a few works have been devoted to graphite oxyfluorides. For such application, most of the studies about covalent GICs concern graphite oxides (GO) and mainly graphite fluorides (GF) [5]–[12]. That is easily explained by high energy density 700 $Wh.kg^{-1}$ (1000 $Wh.l^{-1}$), flat plateau at 2.6 V, broad operating temperature range (–40 to 90°C) and reliability on long term (20 years) recorded for Li/fluorinated carbons (CFx) systems. Nevertheless the rare works on oxyfluorides suggest high potentialities such as an energy density of 1347 Wh.Kg–1 for sample obtained by a two-step synthesis combining fluorination and oxidation [13] and even 2265 $Wh.Kg^{-1}$, thus exceeding graphite fluorides. Fluorinated oxides were highly capacitive while oxidized fluorides have highest discharge potential despite having the same graphite precursor [5].

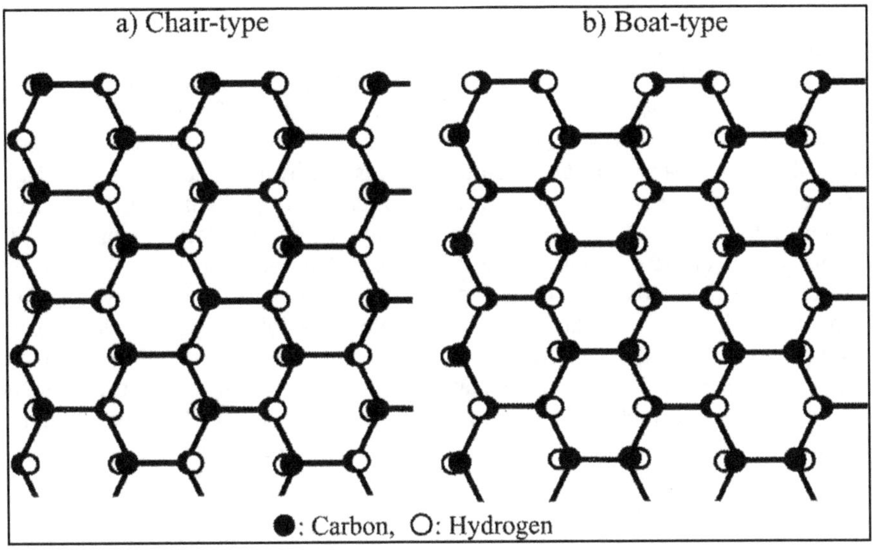

Figure 2: Structure of graphene contains: (a) Chair-type; and (b) Boat-type conformations [4].

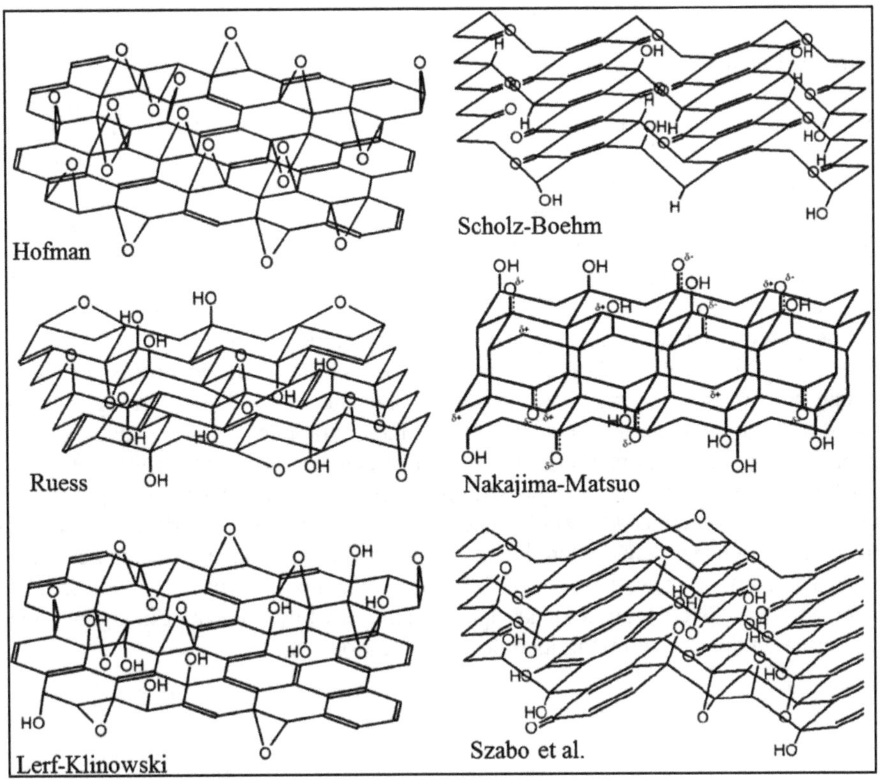

Figure 3: Structure of graphite oxides [4].

2.2.3 Graphene (graphene oxide)

In most cases, graphite oxide was synthesized utilizing the method proposed by Hummers and Offeman54 which was determined from the strategy of Staudenmaire,55 frequently called the Hummers strategy. Its central steps are the oxidation of graphite in concentrated H2SO4 with NaNO3 and KMnO4, the avoidance of overabundance KmnO4 by decreasing to water-soluble MnSO4 with H2O2 and after that washing by methanol. In a few endeavours the Brodie method56 was utilized, where the oxidation of graphite was carried out in smoldering HNO3 with KclO3. To synthesize graphite oxide, the electrochemical oxidation of graphite can be connected in either H2SO4 or HNO3 (conjointly in an alkali arrangement) likely since of the alter in oxygen-containing radicals, but no marked changes within the bulk chemical composition and interlayer dispersing with redundancy of the oxidation prepare. Structure models proposed by distinctive creators are appeared in Fig. 3. Graphite oxide has been utilized as the have fabric for intercalation compounds with distinctive straight natural compounds. Different trimethylammonium cations can be intercalated into the exhibition of graphite oxide with distinctive introductions of long straight particles [4].

2.2.4 Graphene and graphdiyne

MD reenactments were too carried out for three sorts of graphdiynes, considering the same auxiliary parameters as for the graphene i.e., width W~200 Å and tallness L~100 Å and starting arrangement being rolled-up to 4π turns. The structures of the graphdiynes and their 10 comparing nano scrolls are appeared in Fig. 4. The MD recreations have appeared that these nano scrolls moreover come to warm solidness for the temperature considered here (100 K to 1,000 K). The whole recreation time was between 400 ps and 600 ps [6].

2.3 Preparation methods

A few strategies have as of now been set up for creating distinctive sorts of graphene materials. Micromechanical peeling, chemical vapor statement, epitaxial development, circular segment release strategy, intercalation strategies in graphite, unfastening of CNTs, and electrochemical and chemical strategies were a few of the critical planning strategies accessible for the arrangement of graphene. Chemical strategies include solid oxidation of graphite and consequent diminishment to graphene by lessening operators. A novel amalgamation by dichromate oxidation of graphite taken after chemical lessening with hydrazine, which is additionally utilized for the planning of graphene. Electrophoretic statement (EPD) is one of the curiously procedures for synthesizing a nanosheet of graphene. Graphene may well be arranged by coordinate current arc-discharge strategy within the nearness of hydrogen barometrical weight utilizing graphite poles as cathodes for the testimony. Laser pyrolysis method has been illustrated to synthesize multilayer graphene within the nearness of weakening gas. Among these strategies, chemical vapor statement strategies (plasma-enhanced CVD/' generation of graphene materials for distinctive applications. Graphene, delivered in this strategy, was found to have way better crystallinity than that shaped two applications of graphene and graphene-oxide based nanomaterials with any other strategy. PECVD strategy has appeared the flexibility of synthesizing graphene on any substrate, in this way growing its field of applications [8].

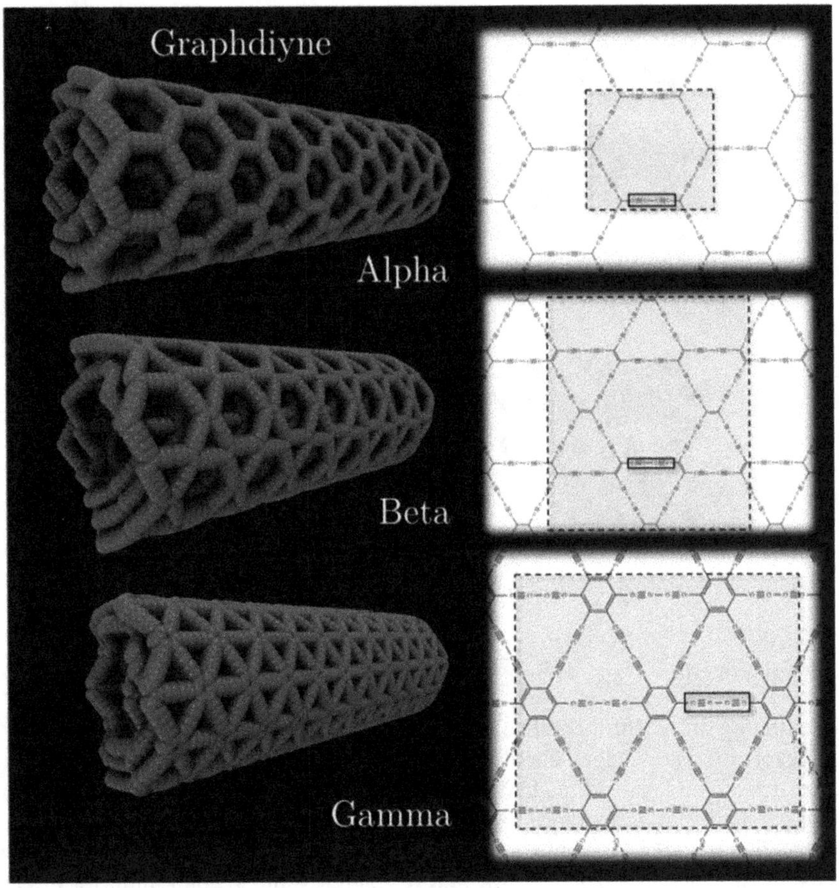

Figure 4: Schematics of the gotten α-, β- and γ-graphdiyne nano scrolls (cleared out side) [6].

2.4 Characteristic properties

The graphene honeycomb cross section comprises of two identical sublattices made of carbon iotas fortified at the side bonds. Each carbon has orbital that contributes to a delocalized organize of electrons. This 3D graphene can have "defects" like topological shapes (pentagons, heptagons, or their combination), edges, opening, adatoms, breaks and adsorbed pollutions [7].

2.4.1 Size, shape, surface morphology and structure

In Fig. 5 schematic outlines of metal oxide–graphene nanocomposites: (a) adsorption of surfactant hemimicelles on the surfaces of the graphene stacks; (b) The self-assembly of anionic sulphonate surfactant on the graphene surface with oppositely charged metal cation (e.g. Sn2+) species and the move into the lamella mesophase towards the arrangement of SnO2 graphene nanocomposites; (c) Metal oxide–graphene layered nanocomposites composed of substituting layers of metal oxide nanocrystals and graphene/graphene stacks; and (d) Self-assembled hexagonal nanostructure of metal oxide antecedent [7].

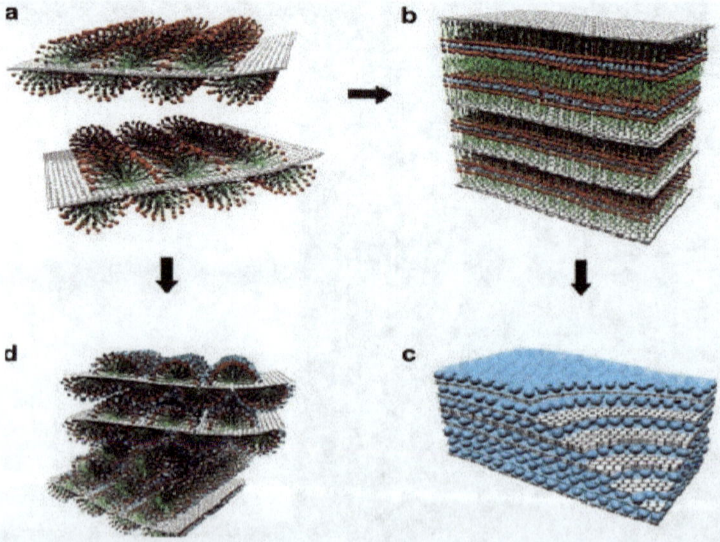

Figure 5: Schematic outlines of metal oxide–graphene nanocomposites [7].

2.5 Thermal properties

By concentrating on recent theoretical and experimental results, we begin by outlining the current level of knowledge about the fundamental understanding of the thermal characteristics of graphene. At the University of California – Riverside, researchers found in 2008 that graphene has an extraordinarily high inherent thermal conductivity K that can be greater than that of carbon nanotubes (CNTs). Excellent thermal characteristics are retained by FLG With an inherent K 2000 W/mK at RT, graphite – the 3D bulk limit for FLG with n layers – remains an exceptional heat conductor. In contrast, K for silver is substantially lower and is 430 W/mK for silver nanoparticles utilised in TIMs. Graphene's heat conductivity was first experimentally studied using a novel non-contact optothermal approach (see Fig. 6) [10].

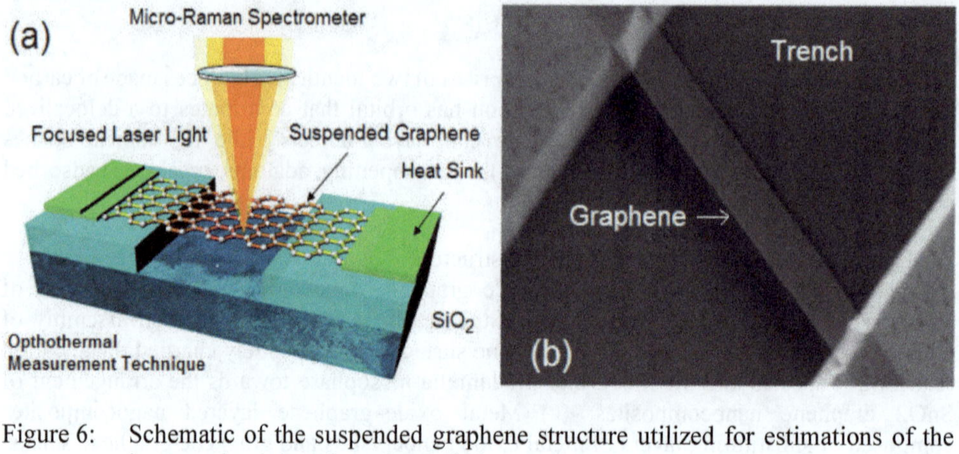

Figure 6: Schematic of the suspended graphene structure utilized for estimations of the heat conductivity of graphene [10].

2.6 Mechanical strength

Another of graphene's stand-out properties is its characteristic quality. Due to the quality of its 0.142 Nm-long carbon bonds, graphene is the most grounded fabric ever found, with an extreme pliable quality of 130,000,000,000 Pascals (or 130 gigapascals), compared to 400,000,000 for A36 basic steel, or 375,700,000 for Aramid (Kevlar) [9].

2.7 Electrical and electrochemical properties

One of the foremost valuable properties of graphene is that it could be a zero-overlap semimetal (with both holes and electrons as charge carriers) with exceptionally high electrical conductivity. Carbon particles have total of six electrons: two within the inward shell and four within the external shell [9].

3 GRAPHENE APPLICATIONS

Graphite and its derivate picked up science and engineering awareness due to its various applications. The disclosure of graphene is properly respected as a point of reference within the world of fabric science; as can be seen within the around the world consideration, the fabric has gotten within the areas of hardware, photonics, capacitors/supercapacitors, biosensing, etc. They are utilized in various applications as outlined underneath. In this book, applications of graphene and its subsidiaries are talked about in detail. These applications incorporate photocatalysis, hardware, gas detecting, graphene-based heterogeneous anodes for vitality capacity gadgets. Fig. 7 shows the different fields that graphene can be used in [11].

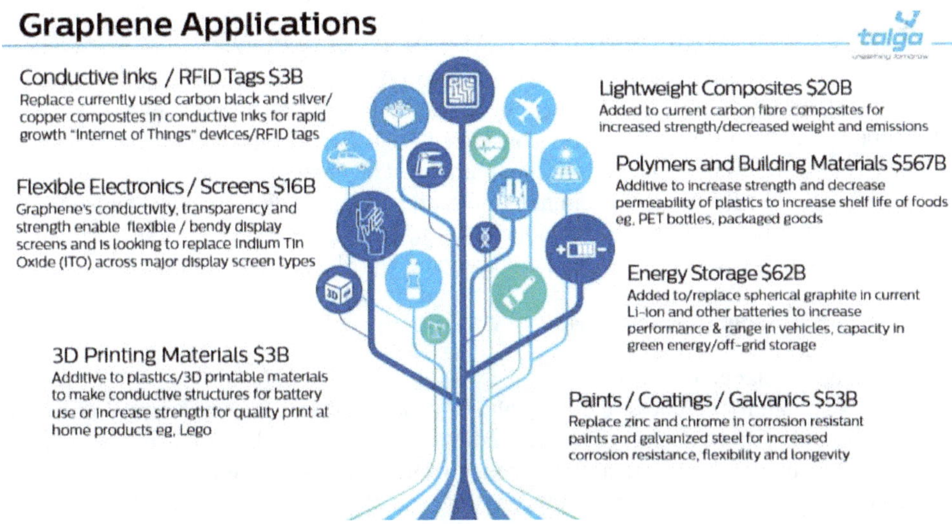

Figure 7: Showcase estimate of future graphene applications. Talga Assets introduction clearly, typically not a versatile prepare. The fetched of creating fair one micrometer of graphene in this way was evaluated in a piece of published in Nature at fair over $1000. Subsequently the name of graphene being the foremost costly fabric within the planet [12].

3.1 Solar cells

Superior get to renewable vitality sources has gotten to be a major concern for creating nations in later times. Sun oriented vitality has been at the cutting edge of nations' endeavors to create modern advances, and researchers are persistently seeking out for modern strategies to create productive energy-generating systems. Graphene, a one-atom-thick sheet of carbon, may be a generally modern fabric that has gotten consideration for its effective capabilities [13].

3.2 Graphene with cement

The strategy utilized by the analysts is known as the high-shear shedding of graphene in water. This strategy permits the water to be supplanted specifically in a concrete blend. The method involves planning graphene suspensions in water by high-shear shedding with the surfactant sodium cholate, coming about in surfactant functionalized graphene (FG). The surfactant functionalized graphene was then mixed in concrete employing a high-shear blender. The test comes about uncovered that the graphene support not as it were upgrades the compressive quality of concrete but too the flexural quality [14].

3.3 Graphene glass

Centered on state-of-the-art design, this development is flexible and versatile arrangement for unused developments or for building recovery. Graphene glass building arrangement are a secluded arrangement that recuperates latticework in veneer, to meet the have to be make strides vitality effectiveness and supportability of buildings [15].

4 GRAPHENE IN ALAMEIN NE NEW CITY

The city is characterized by an expansive number of private units that contrasted in measure and plans to be appropriate for all, but for the normal costs of lofts in it is assessed at 3,700,000 pounds, these costs are considered sensible compared to the level of administrations accessible, and it has been reported that all contracted units will be conveyed as before long as the fundamental administrations and offices for the city are completed, and are anticipated to be completed by the conclusion of 2020 advertisement. A site plan and a 3D simulation for the towers are shown in Figs 8 and 9 [16].

4.1 How to apply grapheme on towers

4.1.1 Graphene solar cell

Graphene's straightforwardness and adaptability make it perfect for sun oriented utilize in spaceflight and air ship, and these properties might make adaptable sun based cell establishment simpler. There are a part of conceivable outcomes when it comes to adjusting graphene to sun powered cells since of its physical and electrical qualities, and its utilization would be greatly valuable to people who expend a parcel of energy. A 3D model simulation is shown in Fig. 10 for solar cells [13].

4.1.2 Graphene glass

The 3D detail shown in Fig. 11 is a modular approach called the Glaphene glass series reimagines latticework. It serves as a space separator or as solar protection for buildings [15].

Figure 8: Site plan for the city which locate the towers to be the most iconic building in Alamein city. *(Source: http://pce-consultants.com/projects/al-alamein-towers-high-rise/.)*

Figure 9: 3D simulation for Alamein towers which contains 40-floor towers comprising of a add up to of 230 units each is involved within the advancement project.

Figure 10: Is a more strong, conducive, and straightforward fabric, it ought to be sent to supplant the ordinary materials utilized in sun based cell [13].

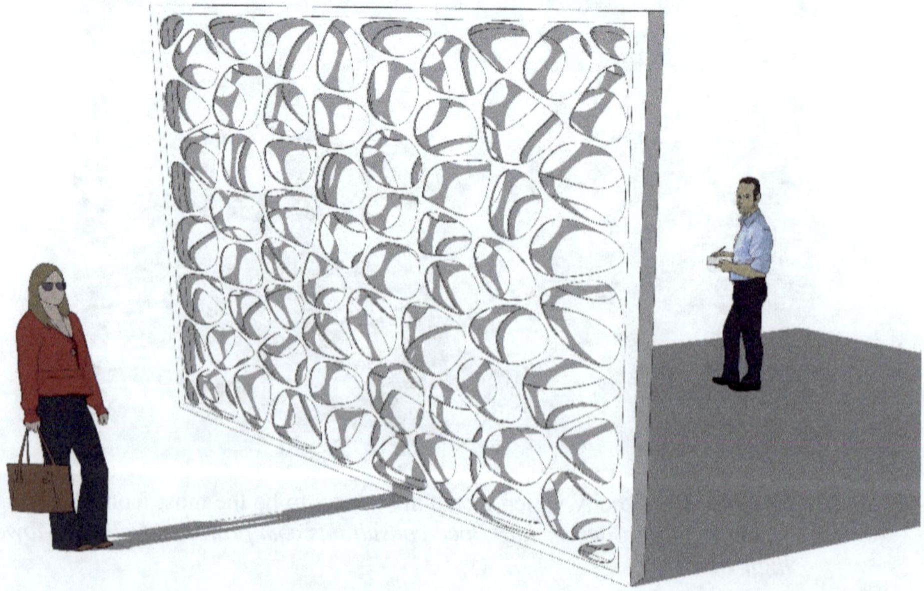

Figure 11: Glaphene glass as an interior solution for the building.

3D detail in Fig. 12 within the tower appear graphene glass settled on façade with its properties that doesn't alter in most extraordinary conditions, it is fire retardant, super hydrophobic and anti-dust [15].

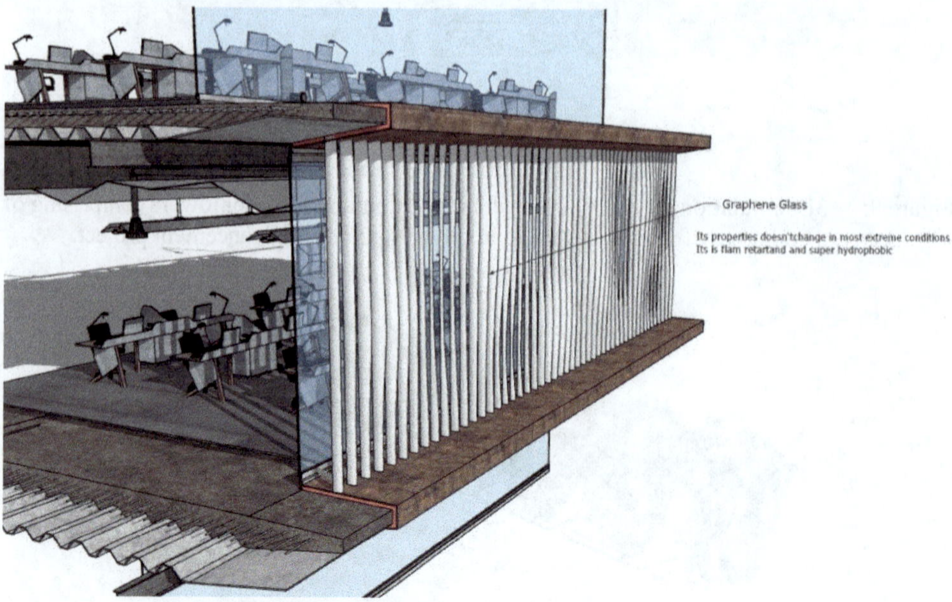

Figure 12: Graphene glass as an exterior solution for the building.

4.1.3 Graphene led light

Adding graphene to an LED bulb helps to dissipate heat – this makes them much brighter, meaning a lower wattage bulb will have the same effect as a traditional LED bulb – effectively reducing energy use for the same resub which is shown in Fig. 13 [17].

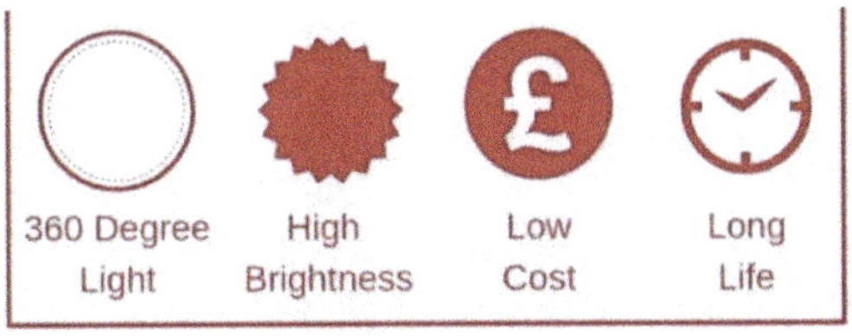

Figure 13: These lights are the most recent innovation within the advancing world of lighting and gadgets, and one step encourage on the way towards cheaper, less harming light bulbs. *(Source: https://www.seratechnologies.com/what-is-a-graphene-led-bulb/.)*

5 CONCLUSIONS

In order to make graphene a more intelligent material, several elements relating to its manufacture, characteristics, and practical applications need to be further explored. Graphene is described as a smart material that is ideally suited for many advanced industrial applications. The potential of graphene, the super substance, is limitless. The possibilities are unlimited because of its extraordinary characteristics and numerous applications.

REFERENCES

[1] Manchester University, Manchester graphene paper among top 100 citations of all time. 2014. https://www.manchester.ac.uk/discover/news/manchester-graphene-paper-among-top-100-citations-of-all-time. Accessed on: Oct. 2014

[2] Raza, H., *Graphene Nanoelectronic*s, pp. 1–12, 2012.

[3] Bradley, D., A chemical history of graphene, *Materials Today*, 10 Jun. 2014. https://www.materialstoday.com/carbon/comment/chemical-history-of-graphene/.

[4] Inagaki, M. & Kang, F., Graphene derivatives: Graphane, fluorographene, graphene oxide, graphyne and graphdiyne, Issue 33, 2014.

[5] Ki, M., Mar, M., Dubois, M., Guérin, K., Batisse, N., Simon, B. & Bernard, P., High energy primary lithium battery using oxidized subfluorinated graphite fluorides. **227**, pp. 1–23, 2019.

[6] Solis, D., Borges, D.D., Woellner, C.F. & Galvão, D.S., Structural and thermal stability of graphyne and graphdiyne nanoscroll structures. 2013. https://arxiv.org/ftp/arxiv/papers/1803/1803.00154.pdf.

[7] Hansora, S.M.D., *Graphene Nanomaterials Fabrication, Properties and Applications*, pp. 1–200, 2018.

[8] Ray, S.C., *Applications of Graphene and Graphene-Oxide Based Nanomaterials*, pp. 11–40, 2015.

[9] de La Fuente, J., Properties of graphene, 2013. https://www.graphenea.com/pages/graphene-propertiese.

[10] Renteria, J.D., Nika, D.L. & Balandine, A.A., Graphene thermal properties. *Applications in Thermal Management and Energy Storage Sci.*, **4**, pp. 525–547, 2014.

[11] Mmaduka-Obodo, R., Ahmad, I. & Ezema, F.I., Introductory chapter. *Graphene and its Applications*, pp. 4–7, 2019.
[12] Su, H., Graphene production beyond the hype: Electrochemical exfoliation, 2015. https://seekingalpha.com/article/3550426-graphene-production-beyond-hype-electrochemical-exfoliation.
[13] Ahmed, U., Graphene: The future of solar cells, 2021. https://www.azom.com/article.aspx?ArticleID=21100.
[14] Ahsan, M.A., Mixing concrete with graphene to build better structures, 2022. https://www.azonano.com/article.aspx?ArticleID=6059.
[15] Vicente Montesinos and Javier Heredia are the founder and executive team Graphenglass, 2020. https://www.graphenglass.com/.
[16] Egypt Real Estate, New Alamein Towers, 2021. https://realestate.eg/en/4738-vacation-in-new-alamein-new-alamein-towers.
[17] Sera Technologies Ltd., https://www.seratechnologies.com/contact-sera-technologies-ltd/.

SECTION 6
DESIGN FOR ECO-ARCHITECTURE

MONOTONY OF PARTS OF CONTEMPORARY ARCHITECTURE: THE ABSENCE OF SMALL DETAILS AND SEPARATIONS

UDO DIETRICH
Resource Efficiency in Architecture and Planning (REAP) Research Group,
HafenCity University Hamburg, Germany

ABSTRACT

In contemporary architecture, a trend can be observed that buildings have a simple rectangular form and façades are composed of a few large elements, all together leading to a monotonous look. Following a proposed list of good design criteria, representative examples of a historical building (from 1900) and a contemporary one are analyzed and confronted. Based on that design criteria an endeavour to criticise certain trends in contemporary architecture can be implemented. To bring the question what makes a façade aesthetic to a more objective basis, the perception mechanism of the human eye is presented. Because of its spherical form, it delivers a sharp vision only in an exceedingly small angle range. Consequently, the eye moves quickly over the object in view producing several sharp spots that are finally completed to a sharp image by the brain. This process can run smoothly if there, where the next eye movement (saccade) strikes on the object, is a new, perceptible element located – the eye's movement is guided, the object is perceived positively and aesthetically. As a precondition, a particularly high number of small-sized elements and few principles of order, what confirms a good part of the list of design criteria, are needed. Based on this, arguments why parts of contemporary architecture may be perceived as monotonous, partly out of human scale and not really aesthetic are formulated as well as conclusions what is missing. In order to involve (future) professionals, a master's course for students of architecture at HafenCity University Hamburg is presented. Architectural ensembles were selected and drawings of the eye's movement scanning the shown objects generated. The students' analysis confirmed well a connection between a good guidance of the eye's movement and a positive, aesthetical perception. Finally, reasons why there is contradiction between the awareness of good design and the practical outcome are sketched.

Keywords: contemporary architecture, visual perception, human eye, saccades, design rules for façades, perception of façades, number, distance and size of elements in well appreciated façades.

1 INTRODUCTION

Criteria for ecological architecture are of course topics like energy and the selection of materials. Nevertheless, buildings and their design frame as well as the public space between them are part of it. A sustainable city doesn't only have buildings with low impact on the environment. It also motivates people to reduce their environmental impact caused by transportation – to slow down and enjoy walking. What makes an attractive walk is, among other things, the pleasure to observe and study façades of buildings. Whether façades are perceived as positive, inviting and finally aesthetic and beautiful is widely subjective and varies from person to person.

The official understanding of urban development in the city of Hamburg (Germany) is, that its typical historical style are brick buildings and that this brand has to be continued also in contemporary architecture.

Fig. 1 shows typical apartment buildings of two different periods, historical and contemporary. The size and arrangement of design elements of the façades as well as the embedding of the buildings in the urban situation can be analysed and confronted (Table 1). This paper deals with the hypothesis, that the criteria listed in Table 1 in the left column are

WIT Transactions on The Built Environment, Vol 210, © 2022 WIT Press
www.witpress.com, ISSN 1743-3509 (on-line)
doi:10.2495/ARC220151

correlated with a positive and aesthetic perception of buildings and their façades. To set it on a more objective basis, Section 2 deals with the operation of the human eye, focussing on how it investigates an object in view along with the brain. Based on these findings, Section 3 is an endeavour to criticise certain trends in contemporary architecture while Section 4 sums up arguments why façades may be perceived as monotonous, partly out of human scale and not entirely aesthetic. Section 5 presents a master's course for students of architecture at HafenCity University Hamburg. Here, eye movement drawings scanning the shown objects should be generated. The analysis by which elements and lines the eye movement is attracted and guided confirms a relation to a positive, aesthetical perception of the object in view – students are aware of what good design is. Section 6 tries to find reasons why only few years later the same persons do not succeed in designing satisfying products, a field for further research.

(a) (b)

Figure 1: Typical apartment buildings in brick style in Hamburg. (a) A historical one, circa 1900; and (b) A contemporary one, circa 2015 [1]. Both buildings are in the same quarter (Barmbek), just a few hundred meters away from each other.

2 EYE AND BRAIN ACTING TOGETHER IN THE VISUAL PERCEPTION

The perception of the human eye is presented here as a short overview, for a more detailed description see Dietrich [2]. The eye is a sphere of a few cm in diameter with one lens in front. The image is opposite of the lens, for physical reasons it can be sharp only in a range of 1° to 2°, the so-called foveal vision (Fig. 2).

Figure 2: The foveal vision of the eye, only here the image is sharp. Only the brain completes it to an extended image (see text) [3].

Table 1: Comparison of size and arrangement of design elements of façades shown in Fig. 1 and the embedding of the buildings in urban situation.

Urban situation		
Transition zone to sidewalk	Yes	No
Entrance door clearly recognizable	Yes	No
Front side of the building is oriented to the street	Yes (balconies and living rooms to the street, even if north oriented)	No, back side (balconies and living rooms to the south oriented internal court)
Architecture		
Upper building end modulated	Strongly modulated with roof and gable	Nearly no modulation, one continuous horizontal line
Use of symmetry	Strong vertical symmetry	No symmetry
Arrangement of design elements in a way that a group of smaller elements form a bigger element	Yes	No formation of groups of smaller elements. Elements are repeated in regular distances but not summed up by design
Façade modulated also in 3D	Yes, strong modulation creates a 3D sight	Weak modulation, rests nearly in 2D-sight
Horizontal and vertical separations	Strong vertical, noticeable horizontal	Nearly imperceptible
Decoration	Supplementary white/green decorative elements, reinforcing vertical separation	No supplementary decorative elements
Use of contrast	High, created by green window frames, white decorative elements and red bricks	Low, dark window frames in dark bricks
Window separation	Symmetrical, down to about 1 x 0.5 m	Asymmetrical, down to about 1.2 x 0.8 m, partly smaller
Size and number of perceptible elements in a way that there are few big elements (formed by groups of smaller ones), middle number of middle sized and high number of small ones	Yes, big elements (in particular vertically), middle sized (windows, balconies) and small ones (decoration, windowpanes, brickwork)	Few big elements, middle number of middle sized (windows), and very low number of small ones (house number, downpipe)
Size of smallest perceptible elements that exist in a higher number	0.2 m (decoration), 0.1 m (brickwork), 1 cm (joints between bricks)	1 m (windowpanes)
Brickwork perceptible	 Clear contrast between red bricks and white mortar	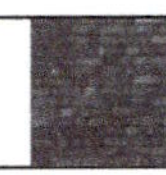 Nearly no contrast, bricks and mortar dark.

The image is more and more out of focus the farther away it is from the foveal view. As a consequence, the eye scans with fast, unconscious movements (saccades) the range around the foveal view. From one saccade to the next one is an angle of 4° to 15°, a movement occurs each 0.2 to 0.6 s [4]. The result is an image composed of several sharp patches. Finally, the brain takes over and adds the missing information out of its experience to a sharp image (Fig. 3).

(a) (b)

Figure 3: Two stages of interaction between the eye and the brain. (a) The eye delivers a patchwork of information by the saccades (similar to the collage of David Hockney, Place Furstenberg Paris, 1985 [5]; and (b) The brain melts it to a complete and sharp image (corresponding to the photography of the place) [6].

If we do not expect something special from the object in view, the movement of saccades is likely haphazardly. However, most often the brain is asking for concrete information, the saccades will be directed accordingly. Conversely, if we look at the object more in a neutral manner, out of curiosity, the object itself will direct the saccades by its offer in details; like we look at a painting, a façade or an architectural ensemble (Fig. 4).

In the case, that there, where the next saccade strikes on, is another, distinct object, the perception is facilitated, guided. The saccades can smoothly explore the object, what will finally be perceived as interesting, balanced, dynamic, and aesthetic.

(a) (b)

Figure 4: (a) View to an architectural ensemble; and (b) The human eye scans the sight with saccades that are attracted and guided by different elements in view [7].

But what is the ideal distance between two elements? Because saccades are done in certain angles, it depends on the distance to the observed object (Table 2). For the most frequent position on the "own" or the opposite sidewalk this distance between two elements should be in the range of several cm to several dm. That means that the elements themselves must have a size of less than half of it to be distinguishable, thus from 4 to 5 cm to few dm. Bigger elements of more than 1 m to a few m complete the set of elements to guide well the observer's eye over the façade.

Table 2: Distance between two nearby objects that are scanned by two consecutive saccades. Saccades occur in angles between 4 and about 15 degrees, the calculated distance depends consequently on the distance to the observed object.

Angle of saccade (°)		*4*	*6*	*8*	*12*	*16*
Distance to observed object (m)						
Immediately in front of	*1*	0.07	0.10	0.14	0.21	0.28
Near on sidewalk	*2*	0.14	0.21	0.28	0.42	0.56
Midst of sidewalk	*3*	0.21	0.31	0.42	0.63	0.84
End of sidewalk	*5*	0.35	0.52	0.70	1.05	1.41
Alley width	*10*	0.70	1.05	1.40	2.10	2.81
Small street width	*15*	1.05	1.57	2.10	3.15	4.22
Street width	*20*	1.40	2.10	2.80	4.20	5.62
Plaza	*30*	2.10	3.14	4.20	6.31	8.43
View from far	*50*	3.49	5.24	6.99	10.51	14.05

And another rule can be derived. To guarantee that there always is an element for the (frequent) small saccades of only a few degrees, the number of the smallest elements in the observed object must be remarkably higher than the number of large elements. A detailed investigation of this "frequency rule" shows, that the necessary number of elements and their size follow a Pareto distribution (Fig. 5).

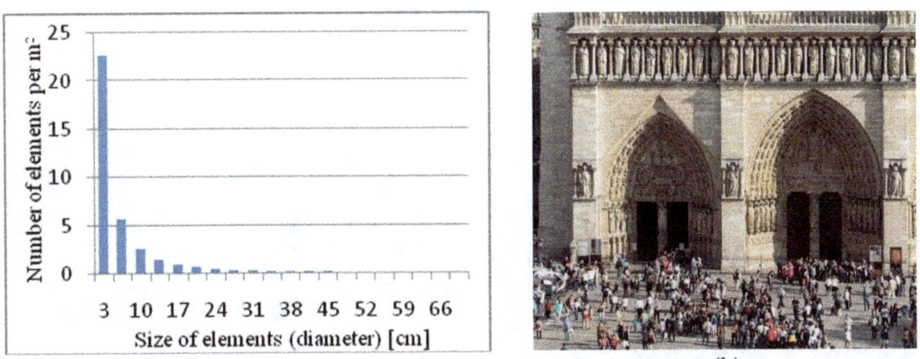

(a) (b)

Figure 5: (a) To satisfy saccades scanning an object, the number of elements must increase with its decreasing size. The correlation follows a Pareto distribution; and (b) The photograph of the portal of Notre Dame in Paris shows an extraordinarily high number of small elements, demonstrating the principle down to dimensions of few decimetre or even centimetre [8].

3 CRITICISM OF CERTAIN TRENDS IN CONTEMPORARY ARCHITECTURE

First it must be mentioned that not all contemporary architecture gives reason to criticise. This investigation refers to a general trend, not to selected buildings. All buildings shown in the figures are just representatives of it and are not the direct object of discussion.

While in other sectors like energy the transition to a sustainable society is meanwhile quite visible, it seems that the design of urban spaces and the corresponding buildings persists still in a variant of a car-oriented philosophy: to deliver quiet conditions, living rooms and balconies and thus the main façades of buildings are oriented off the streets to inner courts. They are felt as semi-public and serve as communication area for a limited house community but not as a real public space. The street side is formed by the backsides of the buildings, leading to a less attractive sight. Supplementary, the ground floors do not show a distinct design and no or hardly offers for retail, leading to a non-attractive atmosphere for a pedestrian strolling along there. Because life is in the inner courts, transition zones to the sidewalks on the street side are not hold as necessary.

Many buildings are of simple shoe-box shape, hardly formed in the upper building line or in the façades that lack a three-dimensional modulation and show a flat, two-dimensional appearance, only balconies (if existent in the façade) often stick out of it.

Further, in regard to the "frequency rule" discussed in Section 2, the arrangement of elements in the façades (especially windows) is not done with the target that a group of smaller elements form a bigger one, two methods are generally used, leading also to a lack of horizontal and vertical separations or axes of symmetry (Fig. 6):

(a) Windows are repeated regularly horizontally and vertically. Windows are of big size and dominate in window to wall ratio, the remaining small parts of the wall between the windows is of the same width horizontally and vertically. Both together form a grid without preferences.
(b) Windows are arranged in a chaotic manner (that demonstrates well that the façade does not belong furthermore to the load bearing system, behind the façade is a concrete structure, the façade can be designed and filled freely).

The number of differentiated perceptible elements is low in all sizes. Especially, the size of the smallest ones is hardly below 0.5 m, the number of these elements is very low (house numbers, down pipes, personal extras of the inhabitants, partly modulation in brickwork).

(a) (b)

Figure 6: Design of façades in a way that a group of smaller elements does not form a bigger element. Realized by arranging windows: (a) In identical size in identical distances; or (b) in a chaotic manner [1].

Contrasts within the façade are widely reduced. Often possibilities to generate a contrast are consciously not applied. The use of dark window frames in a dark brick façade is common, the mortar between the bricks has a dark colour, too (Fig. 7).

(a) (b)

Figure 7: (a) Contrasts in façades reduced by the choice of dark frames in dark brick wall; and (b) White frames and a white plastered staircase form a stronger contrast to the bricks. Both buildings are just opposite each other in the same street [1].

Table 2 shows that for a person that is near to the façade (on the sidewalk) smaller elements would still be necessary to guide the saccades. Besides modulations in the façade and the application of decorative elements the window separation offers the highest potential to create a high number of small sized elements (Fig. 8). From a physical point of view, frames have a worse U-value than present glazing's. As a consequence, a window with many sub separations has higher thermal losses than a window with only few big panes. Advantage and disadvantage should be reflected well before the final design decision – there are also other possibilities to reach a high number of small sized elements.

(a) (b)

Figure 8: Use of window separations to reach a high number of small sized elements. Both are historical buildings, one with (likely) the original window separation (b), delivering a diversity of small elements, also in the brick wall), the other one lost a lot of its character through subsequent interventions; windows were replaced by big panes, supplementary not identical on the ground floor, only the door on the right remembers to the original design (a) [1].

Also, the other part of the façades, the walls between the windows, should show separations and elements. Firstly, they should exist, secondly, they should be distinguishable by contrast or (three dimensional) modulation. Bricks in historical buildings have a red, warm colour; bricks in contemporary façades look darker, browner with an admixture of blue, delivering a cooler sight (Table 1, last row). The colour of the mortar was historically chosen in contrast to the bricks, contemporary often in a similar colour avoiding contrasts and perceptible bricks as smallest elements. Modern industries allow the production of colourless, grey bricks.

Historical brick walls were load bearing, the brick had also the function to act as thermal separation between inside and outside. The thickness of 36 cm bricks and more allowed a highly developed art of brickwork, enriched by further decorative elements (Fig. 8(b)). Contemporary walls need a thermal insulation that exceeds for nearly zero energy buildings 24 cm. Because of physical reasons a full brick layer as outer finish would become too heavy to be fixed through the insulation on the massive wall behind. The last layer is consequently a lighter cladding or a plaster. In case that a brick optic should be kept, thin brick slips or (worse) plastic imitations are glued on the insulation. That could lead to the question if a brick-like sight of contemporary buildings is really consequent or if today's architecture should not be more self-conscious and express that the walls have thermal insulation finished by plaster or cladding.

Finally, it has to be mentioned that the application of thick layers of thermal insulation has the tendency to turn the finish of the wall into a flat, two-dimensional area. It becomes difficult to add here a well three-dimensional modulated sight of the façade – but it is not impossible and is asking for the creativity of planners (Fig. 9).

(a) (b)

Figure 9: Examples of art of brickwork in contemporary architecture. In both cases real bricks were included in the standard brick slips creating a three-dimensional modulation of the façade or accentuating the entrance [1].

4 RÉSUMÉ ON CERTAIN TRENDS IN CONTEMPORARY ARCHITECTURE

Section 2 collected arguments that an object in view like a façade is perceived as positive and aesthetic if the saccades exploring it are easily guided. Accepting that concept, the analysis of Section 3 leads to the conclusion that certain trends in contemporary architecture do not fulfil the described rules. That holds especially for:

- the lack or too low number of small sized elements in the façades (window separations and decorative elements)
- missing or reduced contrasts
- missing or reduced principles of order like symmetry, horizontal and vertical separations, three-dimensional modulation.

Therefore, a saccade starting from one element not striking on another element each time is not guided well. The eye movement scanning the façade will search for a continuation and wander around the motif or even fall out of it. The façade is not observed with pleasure and will not be perceived as positive or aesthetic but rather as monotonous and, in that sense, as out of human scale.

To solve that problem, architecture shall not go back to a historical style! But of course the historical style may serve as a rich source of suggestions for the development of a real modern and human (and in that sense sustainable) architecture. Another source of suggestion could be the knowledge about the processes of human perception. Both together should allow to overcome these trends in contemporary architecture and to pave the way to a human and modern architecture.

5 ARE (FUTURE) ARCHITECTS AWARE OF THESE CERTAIN TRENDS IN ARCHITECTURE?

5.1 Task of a master's course "Human scale and architectural drawing" at HafenCity University Hamburg

Students (of architecture) were asked to have a look at urban architectural ensembles that they like or dislike. The motif should be presented with a photograph. Then, the first task was to create an architectural drawing of it (to study the construction of the perspective). The second task was to observe the motif again and to internalize the corresponding eye movement and finally create a second drawing showing the movement of the saccades (Fig. 10).

(a) (b) (c)

Figure 10: Examples of a students work showing selected (positively perceived) architectural ensembles as: (a) Photograph; (b) Architectural drawings; and (c) Drawing of the eye's movement (saccades) [9].

In the next step students analysed the eye movement and described by which parts of the architecture the eye's view is attracted and how it is moving over the observed object, differentiated according to the positive and negative examples. Further, it should be analysed:

- Number and size of elements, are the smallest elements the most frequent?
- What kind of order single elements in the field of view form (like symmetry, three-dimensional modulation, horizontal and vertical sequences or separations, etc.)?
- What kind of contrasts do exist?

Finally, a statement should be given if the students see a connection between eye movement and aesthetic sensibility when looking at façades, buildings, artworks and if architecture should support the movement of the eye.

5.2 Findings of the students

In total the course had over 100 participants within two semesters, the findings showed that there is a nearly complete compliance between all.

The first finding was that the architectural drawing follows the big, clear lines of the shape of the buildings, big elements of façades, etc. – the drawing shall display the character of the architecture. To general surprise, the drawings of the eye's movement were different. They did nearly not show the big, clear lines but particularly the small and special elements. The saccades continued partly along recognizable principles of order and partly they jumped to the next detail. The guidance of the saccades by different, small elements showed up as something more or less independent of the "frame" of the surrounding big lines.

The positively perceived architectural ensembles showed a big number of small sized elements and several principles of order like symmetry, three-dimensional modulation, a number of smaller elements form a bigger one, horizontal and vertical sequences and separations while the negative ones showed here remarkable deficits. As a criterion of central importance contrasts were recognized, in colour as well as in brightness.

All students confirmed that they see well a connection between a good "guidance" of the eye movement and the positive perception of an object and its contribution to aesthetics. Architectural design should integrate that. Students also expressed that they want to applicate these concepts in their own, future projects.

6 POSSIBLE REASONS FOR THE CONTRADICTION BETWEEN THE AWARENESS OF STUDENTS OF ARCHITECTURE AND THEIR LATER PRODUCTS AS PROFESSIONALS

All in all, it seems that students of architecture are absolutely aware about what aesthetic architectural design is. But why do practical results often show monotonous architecture? What could cause this contradiction?

Architects, as only one part in the group of stakeholders, might be too weak and isolated to fight successfully against these tendencies.

A more sarcastic argument could be that it is very easy to do copy and paste with a design software – et, voila! The building's design is done!

In an interview with a stakeholder of architecture [10] financial reasons are mentioned as responsible. There are so (too) many other cost intensive requirements like energy, fire and sound protection etc. that for cultural aspects like the final aesthetic design is no budget remaining. That statement could have a portion of trueness but also sounds a bit like comfortable excuse. Monotonous architecture is also found in expansive sectors like condominiums (Fig. 6(a), "70 percent sold"). Further, in developed countries societies are richer than all generations before. The former planners had for sure the same problem to construct cheap as possible, but the outcome was often rich and aesthetic – it should be possible today more than ever.

But, in fact, it could well be that they simply deliver what the market is asking for. The students report that they are trained to design buildings that are in first priority different from the others, monumental, representing a status symbol. Buildings are designed isolated of each other, not shaping an urban entity that shows a high grade of diversity at the same time.

….. Present societies in rich countries show a strongly developed individualism. That alone would be a positive fact but in parallel there is a loss in the sense of community. We do not care enough about things and values that we can enjoy only if we act together. The position of space planning and a common sense how to design a quarter or a street has become very weak. At the end it becomes car-oriented because it is common and most simple. Exactly the design of cars expresses an equivalent tendency to a reduced, plump and snobbish design (Fig. 11), confirming that we need a transition in several fields of our society.

(a) (b)

Figure 11: Design of cars. (a) 1970; and (b) 2020. Both are "high-end" versions, held as the most representative and express well the spirit of the time [11].

It seems that the general crisis of climate, resource depletion and overuse, etc., also causes a general loss in cultural understanding. That assertion is related with the way to come out of that crisis and the hope that we are able to realise it and to hand over to the coming generations a planet where it is a pleasure to live on: the transition to a sustainable kind of living.

REFERENCES

[1] Photographs by author, 2020 and 2021.

[2] Dietrich, U., The impact of the human eye's perception on the walkability of street canyons. *WIT Transactions on Ecology and the Environment*, vol. 238, WIT Press: Southampton and Boston, pp. 75–86, 2019.

[3] https://eyetracking.ch/wordpress/wp-content/uploads/2014/01/Gesichtsfeld-Sehbereich-mit-EyeTracking-in-Usability-Tests.jpg. Accessed on: 22 Dec. 2021.

[4] Hasse, C., *Blickbewegungen auf Fassaden*, Shaker: Aachen, 2012.

[5] Hockney, D., Place Furstenberg, Paris, 1985. http://3.bp.blogspot.com/-8MTV9ouBBDI/UQeMNhBo_II/AAAAAAAAWE0/WE3n4vdQF4s/s1600/hockney-furstenberg-paris.jpg. Accessed on: 22 Dec. 2021.
[6] TripAdvisor, Place de Furstenberg. https://www.tripadvisor.de/Attraction_Review-g187147-d191148-Reviews-Place_Furstenburg-Paris_Ile_de_France.html#photos; aggregationId=101&albumid=101&filter=7&ff=320675994. Accessed on: 22 Dec. 2021.
[7] Gilles-Bluhm, T., Human scale and architectural drawing. Master's course at HafenCity University, 2020.
[8] Notre Dame de Paris, La façade occidentale. https://www.notredamedeparis.fr/decouvrir/architecture/la-facade-occidentale/. Accessed on: 3 Jan. 2022.
[9] Dünkel, L., Human scale and architectural drawing. Master's course at HafenCity University, 2020.
[10] Hamburger Abendblatt, Wir dürfen die Städte, die wir lieben, nicht mehr bauen. https://www.abendblatt.de/hamburg/article226157569/Wir-duerfen-die-Staedte-die-wir-lieben-nicht-mehr-bauen.html. Accessed on: 28 Dec. 2021.
[11] Images of cars, in the sequence top left, right, bottom left and right. https://www.autobild.de/klassik/bilder/die-deutschen-autos-der-70er-1298368.html#bild17; https://i.ytimg.com/vi/bqn0SYHtCjU/maxresdefault.jpg; https://i.auto-bild.de/ir_img/7/1/8/7/1/5/Mercedes-300-SEL-6-3-729x486-491e967f8781cfb2.jpg; https://www.motortrend.com/uploads/sites/5/2014/04/Mercedes-Benz-Concept-Coupe-SUV-front-view-posing1.jpg. Accessed on: 29 Dec. 2021.

LANDFORM BUILDINGS AND THE FUTURE OF CONSTRUCTED TOPOGRAPHIES

RENÉ C. DAVIDS
Department of Architecture, College of Environmental Design, University of California, Berkeley, USA

ABSTRACT

Innovative technologies, advanced design methodologies, and an increasing demand for improved building performance prompted by widespread concern for degradation of the natural environment have provoked a reconsideration of architecture's traditional relationship to the ground. In the past, building construction was adjusted as necessary to accommodate challenging terrain, but using advanced modelling technology, contemporary architects now design landform buildings, or landscrapers, which simulate naturally occurring topographic features, including artificial mountains, dramatic geological formations, excavations deep into the earth, vegetated roofs, and landscaped promenades at high elevations walked upon or over. By merging the ground plane and built fabric, new construction can be better integrated with environmentally sensitive sites and reduce resource consumption. Rather than merely follow topographic contours or occupy a given site, landform buildings can provide its replacement. This paper will examine how landform buildings utilize properties usually associated with the natural environment such as continuity with surrounding topography, weathering, contextual integration above and below ground, and programmatic indeterminacy to achieve unique expressions of local, regional, or national identity, improving their surroundings, enhancing their experiential potential and advancing architectural practice.

Keywords: landform building, landform, topography, landscape architecture, constructed landscape, landscraper, environmental architecture, sustainable architecture.

1 INTRODUCTION

In their analysis of what they referred to as the International Style, Henry-Russell Hitchcock and Philip Johnson reinforced historic distinctions between architecture and landscape design, arguing that the natural conditions of the site constitute a pristine entity best left untouched by architects and invoking the myth of the "natural" landscape as the ideal setting for modern buildings: "As far as possible, the original beauties of the site should be preserved. Mere open spaces are not enough for repose; something of the ease and grace of untouched nature is needed as well" [1]. While he did not design underground buildings, the architect Frank Lloyd Wright defined organic architecture as the reinterpretation of nature's principles resulting in more natural forms that better integrated buildings with their surroundings; his Falling Water house (1939) in Mill Run, Pennsylvania is the most celebrated example. Similarly, the work of Alvar Aalto and Elissa Aalto adapted buildings, such as the Säynätsalo Town Hall (1952) to the regional landscape of Finland. In his seminal work *Architecture Without Architects*, Bernard Rudofsky wrote: "The untutored builders in space and time demonstrate an admirable talent for fitting their buildings into the natural surroundings … Instead of trying to 'conquer' nature, as we do, they welcome the vagaries of climate and the challenge of topography … to fit their buildings in"; he could hardly have imagined that rather than fit architecture into the land, some contemporary architects would seek to construct buildings that emerged as topographic replacements [2]. The Earthworks sculpture movement that emerged in the 1960s and gained in popularity and influence throughout the following decade expanded the boundaries of what could be considered sculpture in terms of scale and materials, eventually exerting a strong influence on the environmental arts. Green roofs, artificial mountains, geological forms carved into the earth

WIT Transactions on The Built Environment, Vol 210, © 2022 WIT Press
www.witpress.com, ISSN 1743-3509 (on-line)
doi:10.2495/ARC220161

or emerging from it, buildings walked upon and over, are increasingly prominent in contemporary architecture [3, p. 20].

New technologies and design techniques as well as a demand for enhanced environmental performance have provoked a reconsideration of architecture's traditional relationship to the ground. Some of today's most innovative buildings no longer merely occupy a given site but rather reconstruct the site itself [3, p. 40]. In his book, *Landscrapers: Building with the Land*, critic Aaron Betsky heralded the unfolding of land as the excavation of a visionary and buried utopia [4]. In *Groundwork, Between Landscape and Architecture*, Diana Balmori and Joel Sanders attributed the new way of thinking and working since the turn of the 21st century to "a more nuanced understanding of the complex interplay between human nature and technology" [5, pp. 30–31] that rejects "the false dichotomy between buildings and landscapes" [5, p. 8]. Architects Stan Allen and Mark McQuade examined the many ways in which landscape effects are manifest in architecture today, not as cross disciplinary phenomena involving architects working on the landscape but as "new design techniques, formal strategies, and technical problems within architecture itself" [3, p. 30]. Historian Richard Ingersoll believed a palpable fear that the biosphere was in mortal danger led architects to treat their buildings as landscapes by returning them to the earth and using local materials [6].

A related advantage of buildings which merge with their sites is that substantial energy savings are possible. In a typical above-ground construction, energy is wasted by cycles of heating and cooling, but by reducing heat transfer to and from their surroundings, earth-sheltered buildings require less energy to maintain comfortable living conditions as unwanted infiltration losses are practically eliminated in earth-covered construction [7]. Underground or semi-underground dwellings have provided shelter for millennia in challenging landscapes and climates, often in the form of intricate networks of subterranean dwellings, and more recently for utilitarian buildings such as carparks and shopping malls [8]. Current efforts to reduce energy consumption and the amount of land used for building are motivated by popular concern regarding excessive use of non-renewable raw materials and fossil fuels and the increasing pace of deforestation to clear land for agriculture have increased the focus on the role buildings might play in addressing environmental challenges [9]. As opposed to typical green roofs installed at high elevations, the surfaces of landform buildings can create a physical continuity with their surroundings, making them directly accessible to the public, as well as provide opportunities to incorporate recreational or functional enhancements [10].

2 LANDFORM BUILDINGS

For many contemporary architects, the landscape is no longer a soothing naturalistic background for geometrically pristine buildings [11] but a landscape/architecture hybrid that effectively blends the natural and the synthetic [5, p. 31]. The most important influence on the development of landform buildings has been the availability of new digital technologies that allowed architects to explore the design of fluid organic forms and engineers to calculate complicated stresses while simulating the effects of active forces on natural forms. The Yokohama International Port Terminal designed by the Foreign Office of Architecture (FOA) (2002) is one of the first and probably the most influential landform building to make use of the new technologies, a realization of an architecture invested with landscape techniques operative at the scale of a building (Fig. 1). Yokohama is nothing if not a constructed landscape: the terminal lacks the traditional vertical building walls, substituting a formal language of warped and folded surfaces that suggests an abstract cave-like geology with fluid exterior and interior boundaries, an artificial topography built on a perfectly proportioned podium that could sit anywhere [12].

Figure 1: Yokohama International Port Terminal, FOA. *(Source: Author.)*

Arguably one of the most important tasks of landscape architecture design and by extension landform buildings is to highlight the natural features of a site, a difficult task, as buildings that aim to merge with or replace the landscape forfeit opportunities to establish their identity through subtle contrasts and counterpoint. The Olympic Sculpture Park in Seattle, Washington (2007) by architects Weiss/Manfredi exploits the 40-ft drop in elevation from the top of the site to the water's edge to re-establish the original slope, removed by the installation of transportation infrastructure. The main pedestrian route through the site begins at an exhibition pavilion that integrates with the contours of the park and descends zig zagging as each segment of the path opens to different views of the highways, the Olympic Mountains, the city, port, and a newly created beach, successfully merging architecture, infrastructure, and landscape, with the notable exception of the gallery interior which is not influenced by landscape architecture design [13].

Architect Toyo Ito's Grin Grin Central Park (2005) built on a manmade island in Hakata Bay off the coast of Fukuoka, Japan consists of three greenhouse pavilions with a continuous rippling roof structure of undulating ellipses that appear to be lagoons surrounded by grass. Pedestrian walkways snaking above and below the roof structure create an artificial topography that provides a counterpoint to the flatness of the artificial island. While Toyo Ito's building by being submerged into the ground integrates well with its surrounding landscape, other landform buildings are less successful. Historian William Curtis wrote that Peter Eisenman's City of Culture in Galicia, Spain (2014), promoted for its topographical sensitivity, is an ungainly mass with little relationship to its surroundings that required the complete decapitation of Monte Gaías and the removal of millions of cubic metres of soil from the mountaintop to construct a profile suggestive of rolling hills [14]. Situated on an informal hillside neighbourhood in Medellín, Colombia, the Biblioteca de España (2007) represents a much more successful attempt to merge with the hillside topography. Designed by Colombian architect Giancarlo Mazzanti, the library consists of three interconnected buildings that resemble strange geological monoliths on a hillside platform, each with a different program – library, auditorium, training rooms – covered with dark stone tiles, and perforated with numerous small windows so that they shine like beacons at night [15]. Both Eisenman's City of Culture and the Biblioteca España resemble landforms at a distance, an illusion that disappears on closer inspection. A more easily recognizable landform from far and near is the Oslo Opera House (2007) situated on Oslo's harbourfront, designed by Snøhetta in the form of an iceberg, although its interior is not in any way suggestive of an iceberg [16] (Fig. 2).

Figure 2: Oslo Opera, Snøhetta Architects. *(Source: Author.)*

A more abstract version of a natural form is the Mountain Dwellings project in the Ørestad district of Copenhagen, Denmark (2008) designed by the Bjarke Ingels Group (BIG) in collaboration with JDS Architects (Fig. 3). Stepped dwellings slope towards an adjacent suburb with roof gardens and terraces oriented towards the sun. Built over a parking structure, the apartments can be reached individually by car as if they were suburban houses while pedestrians use an inclined elevator. Two exterior steel stairs also ascend, merge, then climb higher, offering expanded city views, an experience that might be available in a natural setting [17]. While the parking arrangement is spatially spectacular, the building itself is more reminiscent of the clusters of dwellings built on hillsides all along the Mediterranean coast than a mountain.

At the opposite end of the spectrum is the Vanke Centre Horizontal Skyscraper (2009) in Shenzehn, China designed by Steven Holl Architects for a real estate company which while built over a landscaped garden, itself bears no resemblance to a landscape, artificial or otherwise. The building contains compartmentalized programs, including apartments, a hotel, and the head offices of Vanke, grouped together in a linear bar raised on eight white glass columns, so that the ground floor is at the same height as the third floor of most typical commercial buildings, leaving a minimal footprint on the landscape of large-planted mounds, sunken gardens, courtyards, and ponds. While small portions of the program are below grade level and help to give the ground plane its strong presence, the building does not merge with the landscape and has no interior features within that are inspired by it [19]. As was sometimes the case with early Modernist works, the Vanke Centre exemplifies a landscape architecture that does not compete with the buildings but is complementary to the extent that the design would be incomplete without it. By not merging with its surroundings, the Vanke Centre illustrates the difficulty in defining landform buildings as such, and the narrow

Figure 3: Mountain Dwellings, Bjarke Ingels Group. *(Source: Author.)*

expressive range available for landscape practice to inform landform architecture. In many cases the landform is a formal imitation or abstraction of a landscape rather than an attempt to integrate the architecture with nature, thus reinforcing rather than overcoming what Sanders and Balmori describe as the false dichotomy between buildings and landscapes.

3 EXPANDING THE SYNTAX

Among the many essays focused on the relationships between architecture and landscape, one by historian Marc Treib identifies three types: structures that merge, oppose, or synthesize the natural and the man made, yielding a perceptual ambiguity between the identities of each, the precise nature of which is dynamic, seeming to fluctuate under differing environmental or temporal conditions [20]. A linear graphic of possible construction to natural site relationships might indicate merger or synthesis, at least as an aspiration, at one end and opposition at the other. Perhaps the Vanke Centre could be considered a fourth category as it establishes a relationship of affinity and simultaneous distinction between building and an artificial topography.

Within Treib's classification, landform buildings would qualify as structures aiming to merge architecture and landscape, as the default paradigm for landform buildings tend to be natural features such as mounds, hills, and mountains which, if hollowed out, could conceivably become interiors, but exclusive focus on these most conspicuous forms misses the complexity and diversity of other landscapes. Architect Tadao Ando's Chichu Art Museum at Naoshima Island in Japan (2004) has a small although powerful presence on the site's surface as it was built mostly underground to avoid interfering with the beauty of the Naoshima landscape (Fig. 4). It was also a manifestation of Ando's desire to return to what he perceives as the origin of architecture and contemplate light from the realm of darkness with the exclusive use of natural light to illuminate many of the exhibits. The unimposing entrance, just a gash on the hill, leads to a passage that opens to a square courtyard with a view of the sky, planted with long grasses that sway gently in the breeze. A more surprising feature is the concrete passageway leading to the museum's interior that is surrounded by walls tilted at an angle of 6°, according to Ando, "to give visitors the feeling that they are going into the earth, perhaps to its very centre" [21]. A light channel below the massive concrete walls gives the eerie impression that they are floating. The overall structure of the museum consists of an entrance and a gallery wing – the former a square structure and the latter a triangular sunken court – connected by a trench-like channel [22]. The restaurant at the end of the gallery tour offers the museum's only conventional relationship to the exterior, with magnificent views of Naoshima's landscape through a large vertical window. An aerial photograph of the museum shows the geometric shapes of the courtyards and skylights as sculptures in the landscape, appearing disconnected from each other on the surface but linked underground.

Figure 4: Chichu Art Museum, Tadao Ando Architects. *(Source: Mitsuo Matsuoko.)*

The museum was intended to be visited with light as a guide. Ando elaborates that his desire to navigate the relationship between light and dark comes from a personal drive: "I have been looking for my own 'matrix of space', which in my imagination is an obscure place like a cave surrounded by thick, heavy walls of earth, or a space in the darkness lit only with a dim ray of light" [23].

As historian David Leatherbarrow points out, architecture offers more permanent imagery than landscape. Unlike the fixed, geometrically precise apertures of the Chichu Museum, the seasonal cycles of blooming trees, leaves changing colours, and barren branches are familiar spectacles, but in architecture changes of surface colours or textures are seen as problematic, suggestive of deterioration and a compromise of its fundamental role as protection from the elements. An architecture that could be identified as merging with the landscape might use materials that reflect passing of time through processes of decay such as corrosion and oxidation, or the integration of natural phenomena such as water and vegetation, all effects that unless carefully controlled can compromise a building's useful life, but do not necessarily preclude realization [24]. Carlo Scarpa's Querini Sempalia in Venice famously allows water from the canal to flood the building during high tide and light reflected from the water's surface onto its walls and ceilings produces the sort of dramatic change in atmosphere typically associated with landscapes rather than buildings. For Scarpa, water was an unsettling force – a medium mediating between the solidity of the earth and the ever-changing sky. Water is at once reliable and volatile: present like the earth, but in constant flux like the sky. An aqueous quality pervades Querini Stampalia as the floor that wraps up the sides of the wall undermines the sense of solid ground, situating the space within the rhythmic ebb and flow of the tide. While great care is usually taken in architectural design to ensure that no water penetrates a building's protective envelope, water's presence in the landscape generates renewal rather than destruction and Querini Sempalia exists as a sturdy kind of tidepool [25]. The studio by architects Herzog and de Meuron for Swiss Painter Rémy Zaugg in Mulhouse, France allows rainwater mixed with mineral deposits from the building's roof to wash down over its concrete walls, staining them to create a record of the process and as they accumulate, a beautiful abstract painting [26].

4 CASE STUDIES

The more interesting and potent examples of buildings as landscapes are those that find multiple ways to derive inspiration from it. The landscape around Barcelona, Spain has traditionally been regarded as symbolic of its distinctive regional and cultural identity and independence, a focal point for the nation of Catalonia. as well as a source of inspiration for some of its buildings. Located roughly 60 km southwest of Barcelona, the multi-peak mountain range of Montserrat and its extensive honeycomb of caves became the symbol par excellence of the Catalan homeland, exerting an intellectual influence on artists and architects through the Renaixença, a cultural movement promoted by its influential *bourgeoisie* during the last quarter of the nineteenth century: "Catalan mentality was forged in the mountains. We cannot overlook the fact that the human and spiritual resources of this country were to be found in mountainous areas and that the creators of the historical Catalan personality were all mountain men" [27]. In 1900, when industrialist Eusebi Güell commissioned architect Antoni Gaudí, both of whom were Catalan nationalists and devout Catholics, to design a residential suburb on the slopes of the Muntanya Pelada, they drew inspiration from the intricate caves and serrated profiles of the Montserrat mountain peaks [28]. The new development, named Park Güell, was to consist of 60 triangular plots of land for residential buildings that would take advantage of favourable sun orientation and splendid views of the city and sea. As work on the park commenced, prehistoric fossils were found in the numerous small caves during the initial excavation, which later became associated with biblical cataclysms, supporting the notion of scared ground for the new centre of Catalan culture [29, p. 161]. After the development failed to attract buyers and only two houses were built, it became a municipal park in 1923.

The main access to Park Güell leads to a hypostyle hall that was intended to be a market for its residents, with 86 Doric columns supporting a large esplanade situated directly above them that were also used to collect rainwater channelled into a cistern below the market for use in the park. The space between the columns is spanned by domes partitioned with spherical caps. Like a hybrid between an Hellenic temple and a vast cavern, the hall's dark interior contrasts with the expansive open esplanade above it surrounded by a combination of balustrades and benches covered in mosaics [30]. At the summit of the Park Güell hill, negotiated using grotto-like bridges and viaducts surfaced or studded with various rough stones suggestive of the mountain caves of Montserrat, are three rustic stone crosses representing Calvary or Golgotha, the place where Christ was crucified, and within it is a large cavern where iron ore was once mined. The park's paved surfaces as well as its eccentric roof profiles can also be perceived as interpretations of the Montserrat mountain peaks, eroded by wind and rain. The fourth viaduct is perhaps the most evocative of this style, its concave wall set into the ground in the form of an ocean wave supported with inclined buttressing columns, creating a space that suggests a perforated tunnel straddling the interior and the earth's surface (Fig. 5). Gaudí's reverence for the subterranean was influenced by the nineteenth century preoccupation with grottoes, as well as the Catholic revival's fascination with hermit caves [29, p. 147]. Park Güell's bridges resemble both sacred grottoes and primitive buildings, influences apparent in the rough, startingly archaic stonework applied to the columns [29, p. 147]. For some, Gaudi's architecture was a whimsical variant of Art Nouveau; for Nikolaus Pevsner, it was an eccentric Pyrenean interpretation of Medieval and Baroque buildings [31]. Filmmaker Alfred Hitchcock described the park as having "a strange biological plasticity [that] turns structures into malleable masses as in some Gulliverian dream of vegetable or animal elements grown to monumental size" [32]. Beyond its intellectual origins and surreal appearance, Park Güell is like a topographical promenade that winds in and out of the earth, revealing surprising textures and creating formal spectacles up and down the slopes of Muntanya Pelada.

Figure 5: Montserrat Caves (left); and Park Güell Arcade (right). *(Source: Author.)*

A related conceptual voyage featuring excavated areas is the Igualada Cemetery (1994) designed by Enric Miralles and Carme Pinós, the winning design of a competition held in 1985 for a site on the perimeter of the town of Igualada, 65 km west of Barcelona and 20 km west of Montserrat. Situated between an industrial zone and the growing areas surrounding the town, the cemetery is located on a downward sloping site that meets Riera d'Òdena, a stream bordering its east and north sides. After the paved oval entry, a double wall with modular concrete panels creates unique shade patterns to the northeast and hides two unfinished subterranean buildings: a biopsy laboratory that features an imposing, sinuous, curvilinear wall, and a triangular funeral chapel. The triangular form of the chapel has two of its corners truncated to allow access for vehicle parking during funerals. The chapel interior is spare, and lighting is limited to linear incisions in the ceiling, producing a mysterious effect (Fig. 6). While nineteenth and twentieth century Spanish cemeteries such as the Cementerio de la Almudena in Madrid and the previous Igualada Cemetery feature centrally located chapels, the new nearly hidden Igualada Cemetery chapel suggests a return to early Christian catacombs buried in the earth.

Figure 6: Chapel, Igualada Cemetery, Enric Miralles, Carme Pinós. *(Source: Chris DeHenzel, 2012.)*

South of the chapel is an abstract sculpture composed of Corten steel poles resembling the crosses of Calvary at the entrance to a processional pathway. The exterior path of wooden planks embedded in the ground surrounding the main portion of the cemetery is lined with concrete retaining walls containing *loculi*, individual mausoleum-like burial chambers common in Latin countries, capped by a cantilevered half-vault forms that obstruct views of the surrounding landscape; the excavated path is what Dennis Dollens terms a "first-degree boundary", the threshold between a created and existing landscape [33]. The walls resemble the land art sculpture Double Negative (1969), built by earthwork artist Michael Heizer in the Moapa Valley on Mormon Mesa in Utah. The form of the half-vaults is reinforced by the canopies of trees planted on the pathway, creating the appearance of a cylindrical space leading to an oval-shaped terminal plaza that dramatically reveals the sky, surrounded by gabion walls faced with cast iron partitions covering the entrances to family mausoleums that metaphorically represent the final journey into the earth [34]. The metaphors of permanence are represented by excavations into the ground, the gabion stones, the modular concrete paving and niches, tangible, earthly materials juxtaposed with the ever-present reminders of death and of the fragility of human life [35]. Perhaps it was this juxtaposition that led Miralles and Pinós to choose the assertively industrial materials rather than fabricated rock formations as Gaudí had done at Park Güell. The architects did not want to express the perfection implied by machine manufacturing in their work but to mark the effects of the passing of time with aged concrete, rusted Corten steel, cracked wooden railway ties set into the ground, and the assortment of rough stones collected in the gabion walls. These textures, together with the underground spaces, cavern-like passages between levels, sloped walls, and general spatial dynamism are similar both to Park Güell and the Montserrat caves. Miralles and Pinós certainly admired Park Güell and Gaudi's creative manipulation of water, shadows, and movement, as well as his notion of a completely permeable earth and transparent ground. All three architects – Gaudí, Miralles and Pinós – understood that incorporating geological elements into architecture was essential when merging buildings with excavated ground [36].

5 CONCLUSION

While typical landform building are contrived to resemble hollowed out landscape features such as mounds, hills, and mountains, even when successful on its own terms, this approach misses other opportunities to enhance the architectural experience by employing attributes primarily identified with the natural environment such as visible traces of passing time, integration with larger spatial constructs, interconnections above and below ground, programmatic indeterminacy, and the unique, site-specific expressions of local, regional, or national identity. Rather than merely occupy a given plot of earth, innovative landform buildings can provide its replacement, merging ground plane and built fabric, to provide opportunities for recovery of the space typically occupied by buildings alone, as well as reductions in resource consumption. The work of Gaudí and Miralles/Pinos shows that landform buildings can be fully integrated with the landscape without necessarily mimicking its appearance, and the incorporation of a larger range of attributes characteristic of landscape practice can produce a richer, more diversified architectural experience.

REFERENCES

[1] Hitchcock, H.R. & Johnson, P., *The International Style*, W.W. Norton: New York, pp. 76–77, 1966.

[2] Rudofsky, B., *Architecture Without Architects,* Doubleday and Company: Garden City, NY, p. 4, 1964.

[3] Allen, S. & McQuade, M. (eds), *Landform Building: Architecture's New Terrain*, Lars Müller: Zürich, 2011.

[4] Betsky, A., *Landscrapers: Building with the Land,* Thames and Hudson: New York, p. 12, 2002.

[5] Balmori, D. & Sanders, J., *Groundwork, Between Landscape and Architecture*, Monacelli Press: New York, 2011.

[6] Ingersoll, R., *World Architecture: A Cross-Cultural History*, 2nd ed., Oxford University Press: New York, p. 20, 2018.

[7] Bligh, T.P., Shipp, P. & Meixel, G., Energy comparisons and where to insulate earth-sheltered buildings and basements. *Energy*, **5**(5), pp. 451–465, 1980.

[8] Labs, K., The architectural underground. *Underground Space*, **1**, pp. 135–156, 1976.

[9] Bellini, O. & Mocchi, M., Urban green landscapes: The green as contemporary cult. *AGATHÓN International Journal of Architecture, Art and Design*, **2**, pp. 95–101, 2017.

[10] Gissen, D., Architectural reconstruction of nature. *Landform Building: Architecture's New Terrain,* eds S. Allen & M. McQuade, Lars Müller: Zürich, p. 463, 2011.

[11] Constant, C., Social idealism and urban landscape: Sunnyside Gardens versus Romerstadt. *The Modern Architectural Landscape*, University of Minnesota Press: Minneapolis, MN, p. 25, 2012.

[12] Allen, S., Mat urbanism: The thick 2D. *Le Corbusier's Venice Hospital and the Mat Building Revival, (Case Series)*, ed. H. Sarkis, Prestel: New York, pp. 124–125, 2002.

[13] Scalbert, I., Foreign Office Architects' Yokohama International Port Terminal. *AA Files*, **30**, pp. 86–87, 1995.

[14] Curtis, W.J.R., The illusion of plans: Eisenman's landmark vision is a conjuring trick. *The Architectural Review*, **228**, pp. 32–34, 2010.

[15] Broome, B., Giancarlo Mazzanti builds an icon to foster optimism in Medellín, Colombia, with his Parque Biblioteca España. *Architectural Record*, **196**(11), p. 138, 2008.

[16] Davey, P., Grand opera: Opera House Oslo Norway. *Architectural Review,* **223**, pp. 35–45, 2008.

[17] Balık, D. & Allmer, A., This is not a mountain!: Simulation, imitation, and representation in the Mountain Dwellings Project, Copenhagen. *Architectural Research Quarterly*, **March**, pp. 30–40. 2015.

[18] Lebbeus Woods, Steven Holl's horizontal skyscraper. https://lebbeuswoods.wordpress.com/2011/06/28/steven-holls-horizontal-skyscraper/.

[19] Gregory, R., The Vanke Center. *Architectural Review*, **228**, p. 44, 2010.

[20] Treib, M., Inflected landscapes. *Places,* **1**(2), pp. 66–77, 1983.

[21] Ando, T., The Chichu Art Museum. https://www.hatjecantz.de/files/377571460x_06.pdf.

[22] Jodidio, P., *Tadao Ando at Naoshima*. Rizzoli: New York, p. 33, 2006.

[23] Taylor, M.C., *Seeing Silence*, University of Chicago Press: Chicago, IL, p. 249, 2022.

[24] Leatherbarrow, D., Topographical premises. *Journal of Architectural Education,* **57**(3), pp. 70–73, 2004.

[25] Bardt, C., Presence and absence: Carlo Scarpa, Querini Stampalia, and water. https://www.3six0.com/2009/07/23/presence-absence-carlo-scarpa-querini-stampalia-water/.

[26] Arquitectura Viva, Studio Rémy Zaugg, Mulhouse. https://arquitecturaviva.com/works/estudio-remy-zaugg-mulhouse.

[27] Nogué, J. & Vicente, J., Landscape and national identity in Catalonia. *Political Geography*, **23**(2), pp. 113–132, 2004.

[28] Viotti Fernandes, L., Gaudí and the establishment of Catalan cultural identity. https://www.academia.edu/39357533/Gaud%C3%AD_and_the_Establishment_of_Catalan_Cultural_Identity.

[29] Kent, C. & Prindle, D., *Park Güell*, Princeton Architectural Press: New York, 1993.

[30] Sola Morales, I., *Antoni Gaudí*, Harry Abrams: New York, pp. 27–28, 2003.

[31] Pevsner, N., *Pioneers of Modern Design from William Morris to Walter Gropius*, Museum of Modern Art: New York, p. 139, 1949.

[32] Fischer, L., Architecture and the city: Barcelona, Gaudí, and the cinematic imaginary. *Cinema by Design: Art Nouveau, Modernism, and Film History*, Columbia University Press: New York, p. 119, 2017.

[33] Buchanan, P. & Dollens, D.L., The drawing, sitework, and architecture of Enric Miralles and Carme Pinós. *The Architecture of Enric Miralles and Carme Pinós*, SITE/Lumens Books: New York, pp. 6–9, 1990.

[34] Kroll, A., AD classics: Igualada Cemetery/Enric Miralles. https://www.archdaily.com/103839/ad-classics-igualada-cemetery-enric-miralles.

[35] Kogod, L., A commentary on the work of Enric Miralles and Carme Pinós. *Assemblage*, **7**, pp. 108–111, 1988.

[36] Mateo Vega, J.M., Estrategias de Implantación en el Paisaje en Tres Obras de Enric Miralles. Doctoral thesis, Escuela Técnica Superior de Arquitectura de Madrid, pp. 113–173, 2016.

AN INTEGRATED APPROACH TO USER-CENTERED DESIGN: MOBILE BRAIN/HEART IMAGING IN VIRTUAL REALITY

ASMAA BARAKAT, HANY AYAD & ZEYAD EL-SAYAD
Department of Architecture, Faculty of Engineering, Alexandria University, Egypt

ABSTRACT

Poor communication between architects and end-users of the built environment has been implicated as the main barrier to improving design quality and end-user satisfaction. User-centered design (UCD) is an important approach to communication and reaching the end-user affective state. This study aims to develop a new integrated approach for achieving an enhanced user-centered design. UCD is a process of three stages: understanding end-user needs, creating design solutions, and evaluating designs. The new approach is concerned with the built environment's impact on the user's affective state in the early design stage. The mobile brain/heart imaging technology is used to record users' electroencephalogram (EEG) and heart rate variability (HRV) while they navigate the design proposal in non-immersive virtual reality (VR). In this context, the Farsi Lecture Hall in the Faculty of Engineering at Alexandria University was selected to be redesigned. The adopted methodology is based on understanding the needs of architecture students (end-users) through questionnaires and evidence-based design (EBD). Six design proposals were proposed and presented with still images. First, the students were asked to rate each proposal on a 5-point Likert scale. Then they were asked to select the best visualization technique, and the majority chose the 360-panorama. At last, the students sat in Farsi Hall for 3 minutes navigating the best proposal panorama on a full HD laptop screen wearing EEG headband that records EEG and HRV. The SAM test and 5-point Likert stress scale were recorded and compared to EEG and HRV results. The findings of this study show that there is a significant relationship between SAM ratings, EEG, and HRV. Furthermore, 360-panorama is an effective visualization tool for presenting design. Finally, involving mobile brain/heart imaging to induce the end-user affective state while navigating design proposals in VR could enhance design quality and end-user satisfaction.

Keywords: user-centered design (UCD), virtual reality, electroencephalogram (EEG), heart rate variability (HRV), stress, relaxation, emotions, valence, visualization, mobile brain/heart imaging.

1 INTRODUCTION

In recent years, architects and spatial designers have discovered a common interest in how spatial design and the built environment can influence end-user performance and satisfaction. This interest reflects the importance of considering end users' needs and desires in the early design stage, as well as the ongoing demand for new approaches to architecture design that include the end-user in the design process. For decades, understanding user needs and behavior has been regarded as a critical issue that must be considered in product design, giving rise to the term "user-centered design" (UCD) [1]. Many terms for UCD have emerged, including "user/client-oriented design", "people-centered design", "human-centered design", "universal design" and "person-centered design" which are defined as a cycle that starts from the beginning of any design process to its end [2], [3]. UCD is a process of three stages: understanding end-user needs, creating design solutions and evaluating designs [4].

However, UCD appears to have developed and established at a slower rate in the field of architectural design, particularly in large-scale public buildings such as organizations, schools, university buildings, and so on because of two factors. First, designers do not meet with end-users early in the design process. Second, architects (and other key players in the spatial design process such as interior designers, engineers, and others) have traditionally been trained in functional and aesthetic conditions, concerning the various factors that comprise the building as a construction.

In this respect, no clear methodology or framework exists for implementing UCD in public buildings that serve a diverse range of end-users. For example, university buildings, which designed to accommodate and support several academic activities such as teaching, learning, and research. Staff and student productivity, health, and behavior are all affected by the effectiveness of the university spaces. So, university buildings have to provide appropriate facilities and spaces that meet the needs of their end-users (students and staff) [5], [6]. Similarly, lecture halls are an integral part of the physical learning environment in the university and can impact the entire learning process, as well as students' well-being, comfort, and productivity. The overall performance of a lecture hall is determined by specific parameters [7].

This study aims to develop a new integrated approach for achieving enhanced user-centered design in designing and retrofitting spaces in public buildings to meet the needs of end-users. The new integrated approach is concerned with the built environment's impact on the user's affective state in the early design stage. To accomplish this aim, we employed two affective state measures: self-report measures and physiologic measurements (mobile brain/heart imaging). Following that, we compared the two measures to determine the best method for evaluating design proposals and quantifying end-users affective states during the early design stage.

The first part of this paper (theoretical study) investigates the evidence-based design of the educational rooms in general. Particularly, this paper focuses on four design elements: colors, lighting, biophilic design, and furniture. Affective state measures, such as self-report and physiologic measurements, are also reviewed. The second part of this paper (empirical study) discusses the adopted methodology which is based on proposing six design proposals for the Farsi Lecture Hall and asking the students (end-users) to select the optimum proposal. At last, the students sat in Farsi Hall for 3 minutes navigating the best proposal panorama on a full HD laptop screen wearing a mobile brain/heart headband that records EEG and HRV and filled self-report questionnaire. The results of this study depict that there is a significant relationship between self-reporting, EEG, and HRV. Furthermore, 360-panorama is an effective visualization tool for presenting design. Finally, involving mobile brain/heart imaging to induce the end-user affective state while navigating design proposals in VR could enhance design quality and end-user satisfaction.

2 LITERATURE REVIEW

2.1 Evidence-based design (EBD)

A long-standing challenge in lecture hall design is how to influence the built environment to improve cognitive performance, affective state, relaxation, and stress reduction. As a result, the literature on stress and relaxation in the built environment was reviewed. A study was conducted to determine the effect of classroom wall colors on student attention. Purple wall color conditions were found to have the highest levels of attention, followed by blue, green, yellow, and red wall colors [8]. Similarly, according to another study, students preferred the colors yellow, dark blue, red, and light blue in that order [9]. Later, a study found that blue is the most popular color among students, followed by green, violet, orange, yellow, and red [10]. As a result, for our case study, we used light blue for the walls and dark blue and yellow for the furniture.

In terms of lighting, previous research has shown that using the standard lighting setting which is characterized by average horizontal illuminance measured at desk level is 300 lx, the CCT is 3000–4000 K, and the standard white light can improve concentration and alertness, and increasing productivity [11]. Furthermore, higher correlated color temperature LED lighting has a perceived positive impact on behaviors and emotional/mood responses, with bluer/higher kelvin temperatures improving alertness, focus, and arousal [12]. As a result, we chose LED lighting for our case study; the illuminance measured at the desk level is 300 lx, the CCT is 3000–4000 K, and the light is standard white.

According to the reviewed literature, students benefit from biophilic design in terms of cognitive function, emotional/mood responses, relaxation, and stress reduction [13], [14]. There are many patterns of biophilic design. We chose two major biophilic concepts for our research: nature in space patterns and natural analog patterns. Nature in space can be represented in visual connection with nature, plants, and dynamic and diffused lighting. Meanwhile, natural analog patterns can be represented in finishes, and objects with biomorphic and fractal forms [15].

Concerning furnishing, student responses gathered during the event "What makes a classroom great?" held as part of the transforming the instructional landscape (TIL) as a major classroom redesign at the University of Toronto, revealed that the FT20 chair was strongly preferred. The FT20 was popular because of its cushioning and attached desk [16]. As a result, we used the FT20 chair for our case study.

2.2 Affective state evaluation

The process of measuring a user's specific emotions is known as affective state evaluation. It has been extensively researched, and numerous techniques have been developed. There are various types of evaluation; however, for this study, we only used self-report measures and physiological measures (mobile brain/heart imaging).

2.2.1 Self-reporting measure

Self-reporting necessitates the user verbally or nonverbally reporting his or her feelings. The only way to measure the first component of emotion: a person's subjective feeling, is through self-reporting. This study will employ various self-reporting techniques, including the Patient Health Questionnaire-9 (PHQ-9), the 5-point Likert scale, and the Self-Assessment Manikin (SAM). First, the PHQ-9 is the depression module, in which scores of 5, 10, 15, and 20 indicated mild, moderate, moderately severe, and severe depression, respectively. Second, the 5-point Likert scale was chosen as the primary instrument in all stages of this study because it is the simplest and most practical way to measure the strength of satisfaction and stress [17]. Finally, we used the SAM which has two dimensions, valence (negative/positive emotions) and arousal or activation (low/high), and every emotion can be placed on these two-dimension of SAM. For example, "fear" would be an emotion with a high level of activation and negative valence, while "calm" would be a low-activation positive-valence emotion. Furthermore, high arousal and positive valence are associated with feelings of happiness, excitement, and alertness. Conversely, low arousal and positive valence translate as calm, relaxed, content, and serene [18].

2.2.2 Mobile brain/heart imaging (physiological measure)

The most commonly used physiological measures of affective state are autonomic nervous system indicators, such as heart rate variability (HRV), and electrodermal measures, such as electroencephalography (EEG). We chose a biometric sensor that can measure EEG and

HRV after considering several factors. First, we decided to use a simpler single-electrode EEG headband because it is easier to manage, smaller, participants are not afraid to use it, and it could be compatible with a VR headset if it is used [19]. Second, we plan to use a consumer EEG headset that will be widely used in the future, and searchers and architects are unconcerned about high costs. Third, we preferred an EEG headset that produces real-time affective states such as stress and relaxation, so that searchers and architects could use it without the need for a programming team or complicated analysis. Furthermore, the literature review supported our considerations by revealing that low-cost EEG headbands can be used to detect emotion in virtual reality [20], [21].

For our experiment, we chose the Flowtime Biosensing Headband which measures EEG and HRV. The output raw data from the Flowtime headband includes EEG absolute brainwave rhythm (dB) and power ratios (alpha α, beta β, delta δ, Theta θ, Gamma γ), heart rate variability (HRV), attention, relaxation, and stress levels. The higher the alpha and theta waves, the greater the relaxation level. While beta and gamma waves are associated with being active and thinking or when a person is highly focused during meditation. On other hand, delta wave refers to deep sleep. Furthermore, the higher the HRV, the greater the relaxation. We tested the Flowtime headband on a 30-year-old male participant who wore the Flowtime headband and watched IAPS emotional pictures of mirth, disgust, and fear, as well as different videos of natural scenes and slums without any sound. We discovered that EEG oscillations, HRV, stress, and relaxation levels varied depending on what was watched.

3 METHODS AND TOOLS

In order to achieve the previously mentioned aim, developing a new integrated approach for achieving enhanced user-centered design in designing and retrofitting spaces in public buildings to meet the needs of end-users, the following steps were adopted (Fig. 1):

1. Understanding the needs of end-users through questionnaires and evidence-based design.
2. Creating various design proposals.
3. Evaluation of the proposals to select the optimum proposal.
4. Using Mobile Brain/Heart Imaging and self-reporting measures to quantify the optimal proposal's end-user affective state.

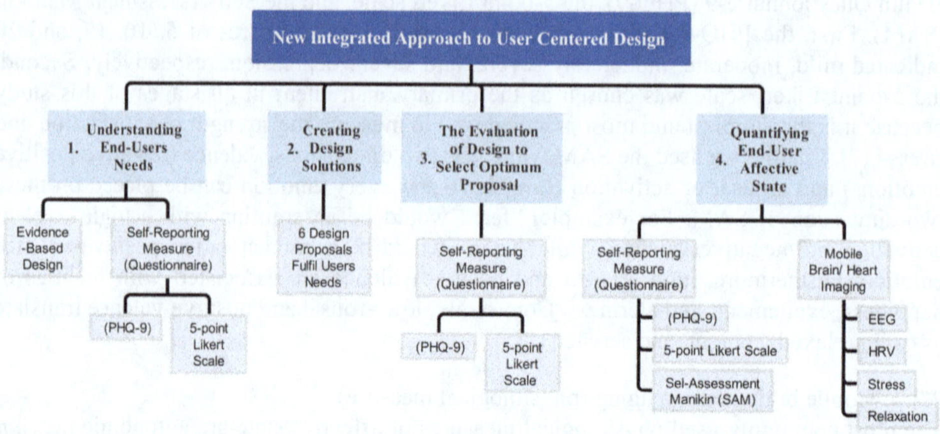

Figure 1: The integrated new approach to the UCD framework.

Our vision is to integrate biometric sensors into non-immersive virtual reality to measure EEG and HRV responses. Besides integrating self-reporting measures such as the SAM and a 5-point Likert scale to evaluate how users feel while interacting with the design proposal in non-immersive VR. Finally, statistical methods are used to quantify the sensor raw data captured throughout the experiment, as well as SAM and the 5-point Likert scale of stress results. The scope of this paper is limited to the affective states of stress, and relaxation in VR, and the biometric sensor data captured and analyzed are EEG and HRV only.

3.1 Study location: Farsi Lecture Hall

This study was conducted in the Farsi Lecture Hall at Alexandria University's Faculty of Engineering to be redesigned to meet the needs of students while adhering to UCD principles. The Farsi Hall is a large, traditional classroom that can accommodate 100–150 students. The Farsi Hall is distinguished by wood-clad walls, parquet floors, a white ceiling, and artistry drawings on the upper part of the walls. Tiered seating is provided, with fixed bench tables and movable leather chairs as illustrated in Fig. 2.

(a) **(b)**

Figure 2: (a) The real Farsi Hall perspective; and (b) Farsi Hall plan.

3.2 End users' needs (questionnaire)

In this phase, we randomly distributed and collected 65 questionnaires. Among them, 60 (92.3%) questionnaires were valid. Participants included 12 male and 48 female architecture students, who were randomly chosen from the department of architecture in the Faculty of Engineering at Alexandria University. All students were aged between 21 and 25 years old. The students' educational backgrounds varied, with 11 being postgraduate students and 49 being undergraduates. The questionnaire was carried out in Farsi Hall. It began with demographic information and was followed by the PHQ-9. Five participants scored > 5 in PHQ-9, and they were all eliminated. The questionnaire then presented a 5-point Likert scale of satisfaction to determine the satisfaction level with the Farsi interior component such as color, textures, walls, ceiling, lighting, furnishing, and environmental aesthetics. Finally, the question of whether biophilic design should be integrated into Farsi Hall was raised.

3.3 Creating design proposals

Six design proposals were proposed based on end-users' needs questionnaire results and evidence-based design. All of the proposals follow the biophilic design. They all share the principle of nature in space, which is represented by plants outside on the window wall "The visual connection to nature" and motorized, perforated, and translucent roller blinds that print window plant shadows "Dynamic and diffused lighting" (Fig. 3(a)). Furthermore, they all adhere to biomorphic forms and patterns achieved by graphic abstract leaves covering the board wall, floor carpet with prairie grass print, front table made of 3D vertical tiles carved into the shape of a wave, and board wall wood cladding material. In terms of furniture, they all use the FT20 chairs with blue cushioning and a wood back (Fig. 3(b)).

Figure 3: (a) Translucent window blinds and outside plants; and (b) Graphic leaves wall cover, prairie grass print on floor carpet with, 3D carved waves front table, and board wall wood cladding.

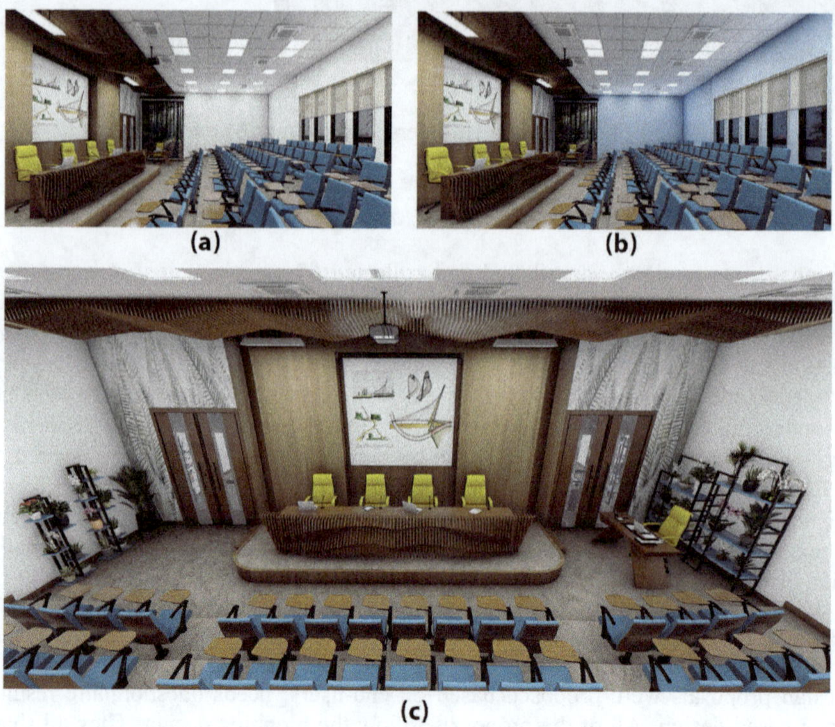

Figure 4: (a) Proposal 1; (b) Proposal 2; and (c) Proposal 3.

Moreover, the LED lighting illuminance at the desk is 300 lx, the CCT is 3000–4000 K, and the standard white light is approved for all proposals. Whereas the walls of proposals 1, 3, 5, and 6 are white (RAL 9010), and the walls of proposals 2 and 4 are blue (RAL 5012). Indoor 19 plants were selected from the guidance set out by Wolverton in his publication "How to grow fresh air" [22] are used in proposals 3, 4, 5, and 6) while the mural photographs of woodland on the two-sided walls in the front part of the hall are applied in proposals 1, 2, and 6 as Figs 4 and 5. Finally, wood cladding of two-sided wall and ceiling is used in proposals 5 and 6 as indicated in Fig. 5.

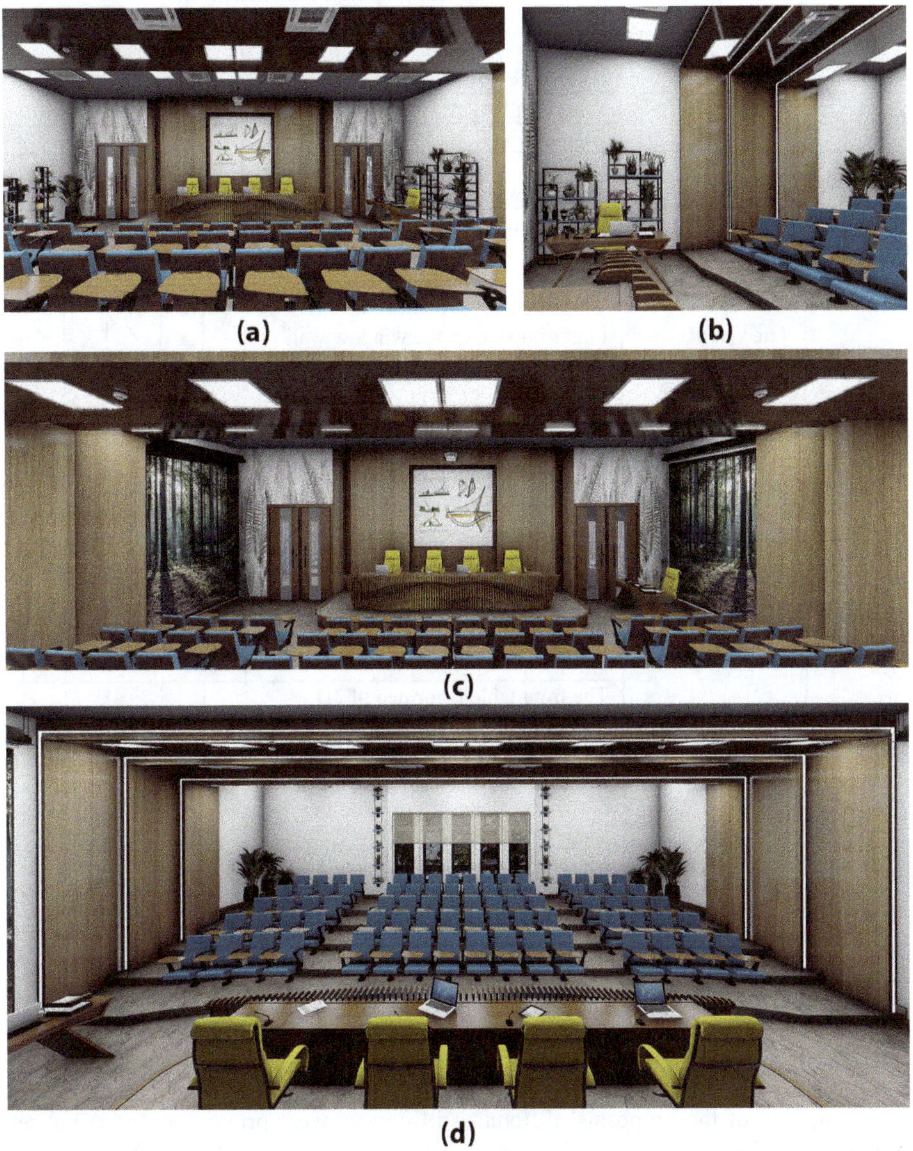

Figure 5: (a) Proposal 5 front elevation; (b) Proposal 5 side elevation; (c) Proposal 6 front elevation; and (d) Proposal 6 back elevation.

Proposal 1 is identical to proposal 2, except that the walls in the proposal 1 are white and the walls in proposal 2 are blue. Proposals 3 and 4 are the same, with the proposal 3 in white and proposal 4 in blue. Meanwhile, proposal 5 is identical to the proposal 6 in that both use wood cladding on two-sided walls and ceilings, with the exception that proposal 6 uses mural photographs of woodland and indoor plants in the back, whereas proposal 5 only employs indoor plants. All the previous design considerations are presented in Table 1 and Figs 3 and 4.

Table 1: Overview of proposals' design considerations.

<table>
<tr><th colspan="3" rowspan="2">Design considerations</th><th rowspan="2">Elements</th><th colspan="6">Proposals</th></tr>
<tr><th>1</th><th>2</th><th>3</th><th>4</th><th>5</th><th>6</th></tr>
<tr><td colspan="3" rowspan="2">Wall colors</td><td>White walls (RAL 9010)</td><td>✓</td><td>×</td><td>✓</td><td>×</td><td>✓</td><td>✓</td></tr>
<tr><td>Blue Walls (RAL 5012)</td><td>×</td><td>✓</td><td>×</td><td>✓</td><td>×</td><td>×</td></tr>
<tr><td colspan="3">Furnishing</td><td>FT20 chairs blue cushioning and wood back</td><td>✓</td><td>✓</td><td>✓</td><td>✓</td><td>✓</td><td>✓</td></tr>
<tr><td colspan="3">Artificial lighting</td><td>LED lighting illuminance measured at desk level is 300 lx, the CCT is 3000–4000 K, standard white light</td><td>✓</td><td>✓</td><td>✓</td><td>✓</td><td>✓</td><td>✓</td></tr>
<tr><td rowspan="10">Biophilic design considerations</td><td rowspan="3">Nature in space</td><td rowspan="2">The visual connection to nature</td><td>Plants outside on the window wall</td><td>✓</td><td>✓</td><td>✓</td><td>✓</td><td>✓</td><td>✓</td></tr>
<tr><td>Indoor plants</td><td>×</td><td>×</td><td>✓</td><td>✓</td><td>✓</td><td>✓</td></tr>
<tr><td>Dynamic lighting</td><td>Roller blinds are motorized, perforated, and translucent to print window plant shadows</td><td>✓</td><td>✓</td><td>✓</td><td>✓</td><td>✓</td><td>✓</td></tr>
<tr><td rowspan="7">Natural analogs patterns</td><td rowspan="5">Biomorphic forms and patterns</td><td>Graphic abstract leaves cover the board wall</td><td>✓</td><td>✓</td><td>✓</td><td>✓</td><td>✓</td><td>✓</td></tr>
<tr><td>Floor carpet with the print of prairie grass</td><td>✓</td><td>✓</td><td>✓</td><td>✓</td><td>✓</td><td>✓</td></tr>
<tr><td>The front table is formed of 3D vertical tiles carved into the shape of a wave</td><td>✓</td><td>✓</td><td>✓</td><td>✓</td><td>✓</td><td>✓</td></tr>
<tr><td>3D ceiling tiles are made up of horizontal planes carved into the shape of a wave</td><td>✓</td><td>✓</td><td>✓</td><td>✓</td><td>×</td><td>×</td></tr>
<tr><td>Mural photograph of woodland on the two-sided walls in the front part of the hall</td><td>✓</td><td>✓</td><td>×</td><td>×</td><td>×</td><td>✓</td></tr>
<tr><td rowspan="2">Materials connected to nature</td><td>Wood cladding of board wall</td><td>✓</td><td>✓</td><td>✓</td><td>✓</td><td>✓</td><td>✓</td></tr>
<tr><td>Wood cladding of two-sided wall and ceiling</td><td>×</td><td>×</td><td>×</td><td>×</td><td>✓</td><td>✓</td></tr>
</table>

3.4 Proposals evaluation to select the optimum proposal

At this stage, all of the proposals' different still images were printed on one A0 sheet to facilitate comparison and hung in the foyer in front of Farsi Hall. We randomly distributed and collected 75 questionnaires. Eleven participants scored > 5 in PHQ-9, and they were all eliminated. 64 participants included 15 male and 49 female architecture students, who were randomly chosen from the Department of Architecture in the Faculty of Engineering as Fig. 6. All students were aged between 21 and 25 years old. The questionnaire began with gender and age questions, followed by a 5-point Likert scale of satisfaction to determine the level of satisfaction with each proposal. Then comes the question of which proposal is the most favorable. Finally, we inquire about the most preferred visualization technique (still image, 360-panorama, walkthrough video, and so on).

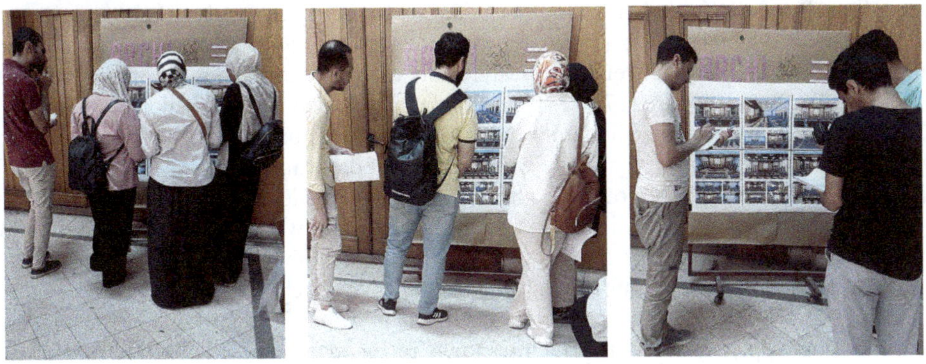

Figure 6: Proposals evaluation to select the optimum proposal.

3.5 Mobile brain/heart imaging and affective state

Based on the previous stage's results, proposal 6 was selected as the optimum proposal and presented as a 360-panorama on a full HD laptop wearing EEG Headband that records EEG and HRV. The laptop screen was chosen based on the findings of the study that discovered computer screen visualization (non-immersive VR) maintains a higher perceived quality than an immersive environment [23]. In the first, the PHQ-9 questionnaire was randomly distributed and collected from 42 students from the department of architecture in the Faculty of Engineering. Among them, six participants scored > 5 and they were eliminated. The 36 students who completed the experiment ranged in age from 21 to 25 years old. They spent 3 minutes in Farsi Hall navigating a 360-panorama of the best proposal 6 on a full HD laptop screen while wearing a Flowtime headband that recorded EEG and HRV Fig. 7. The SAM test and a 5-point Likert scale of stress level were recorded and compared to the results of the EEG and HRV.

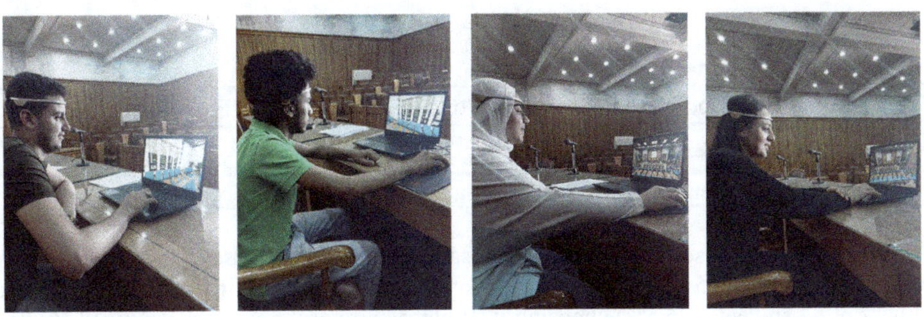

Figure 7: Participants navigating proposal 6 panorama wearing EEG headband.

4 RESULTS

The data analysis results from the experiment stages are summarized below.

4.1 End-users' needs questionnaire

IBM SPSS Statistics v26 and MS Excel were used to generate descriptive statistics of the end-users' needs questionnaire. Total scores, distribution of percentages, and weighted mean were all analyzed. Weighted means were calculated for the Likert scales, ranging from very unsatisfied = 1 to very satisfied = 5, to determine the tendency of the composite scores. Based on the reviewed studies, the weighted mean results can be interpreted to show how satisfied (or dissatisfied) students are [17], [24]. The findings revealed that the majority of students, 42 of 60 (70%), are very satisfied with the artistic drawings on the upper part of the walls, with a weight mean score of 4.52. The wood texture of the walls and floor received the next highest weighted mean score of 3.47, indicating that the students are satisfied with it. The weighted mean score for artificial and natural lighting was 3.37, indicating neutral. Furnishing is also neutral, with a weighted mean of 3.1. On the Other hand, 60% of students were dissatisfied with environmental aesthetics, and 28.3% chose neutral with a weighted mean of 2.25 to represent dissatisfaction as shown in Table 2. Finally, when the students inquired about applying biophilia principles to the redesign process, 36 of 60 (60%) students strongly agree, 16 (26.7%) agree, 2 (3.3%) chose neutral, and 6 (10%) disagree. The weighted mean of applying biophilia principles is 4.3, indicating a strong agreement.

Table 2: The end users' needs questionnaire result.

Farsi components	Artistic drawings	Wood texture (walls/floors)	Lighting	Furnishing	Aesthetics
Weighted mean	4.52	3.47	3.37	3.1	2.25
Result interpretation	Very satisfied	Satisfied	Neutral	Neutral	Unsatisfied

4.2 The optimum proposal questionnaire

The optimum proposal questionnaire data were also analyzed using IBM SPSS Statistics v26 and MS Excel. Weighted means were calculated for the Likert scales, ranging from very unsatisfied = 1 to very satisfied = 5, to determine the tendency of the composite scores. According to the findings, the weighted mean of the proposals 5, 6, and 3 indicates that the students are satisfied. Furthermore, the weighted mean of the proposals 1, 2, and 4 indicates neutral as shown in Table 3.

Table 3: Proposals evaluation results.

Proposals	Proposal 1	Proposal 2	Proposal 3	Proposal 4	Proposal 5	Proposal 6
Weighted mean	3.34	2.77	3.89	2.95	4.06	4.06
Interpretation	Neutral	Neutral	Satisfied	Neutral	Satisfied	Satisfied

According to the results of the best proposal question, the proposal 6 is preferred by the majority of students 25 of 64 (39.1%). Proposal 3 was the second most popular proposal, with 17 of 64 students (26.5%) choosing it. Whereas proposal 5 was chosen by 13 students (20.3%). On the other hand, proposal 1 was chosen by five students (7.8%), proposal 2 by one student (1.6%), and proposal 4 by three students (4.7%). As a result, proposal 6 is regarded as the best proposal for the end-users. Proposal 6 is characterized by using mural photographs of woodland, indoor plants in the back, and wood cladding of two-sided wall and ceiling as illustrated in Fig. 8. Finally, 43 out of 64 students (67%) chose 360-panorama as the best visualization technique.

Figure 8: Proposal 6 which is selected as the optimum proposal.

4.3 The results of mobile brain/heart imaging stage

According to SAM results for the proposal 6 design, 15 of 36 students (41.7%) chose the highest level of valence (very positive emotions), while 18 of 36 students (50%) chose the second positive degree of valence (positive emotions), and three students (8.3%) selected neutral, while there is no one selected negative emotions. In terms of arousal, two students (5.6%) chose the highest level of arousal (very high arousal), eight students (22.2%) chose the second-highest level of arousal (high arousal), six students (16.7%) chose neutral, 10 students (27.8%) chose low arousal, and 10 students (27.8%) chose very low arousal. On a 5-point Likert scale of stress level, 50% of students chose “no stress”, 27.8% chose “low stress”, and 22.2% chose “neutral”. The raw data from the Flowtime headband is then analyzed, and their weighted means are calculated and compared to the weighted means of SAM and 5-point Likert scale stress level results.

5 DISCUSSION

The findings revealed that when participants in the SAM test have high arousal and positive valence (meaning high excitement and low relaxation), HRV, relaxation, alpha α-power ratio, and beta β-power ratio are all low, but they are all higher when participants have low arousal and positive valence (meaning high calm relaxation) as shown in Fig. 9. These results are associated with literature. But there is no relationship between these results and theta θ, gamma γ power ratios. Besides, there is a clear link between weighted means of 5-point Likert scale stress and headband calculations of HRV, stress, and relaxation. According to the literature, HRV is high when a person is in a relaxed state; HRV is high in participants who said they have no stress and decreases in participants who said they have low and moderate stress, respectively. Moreover, weighted means of 5-point Likert scale stress and weighted means of calculated headband stress have a directly proportional relationship. while weighted means of 5-point Likert scale stress and weighted means of calculated headband relaxation have an inverse proportion relationship as illustrated in Fig. 10. On the other hand, there is no relationship between 5-point Likert scale stress and the analysis of alpha α, beta β, delta δ, theta θ, and gamma γ raw data. As a result, the Flowtime headband is effective in stress interpretation by relying on its app equations of HRV, relaxation, and stress.

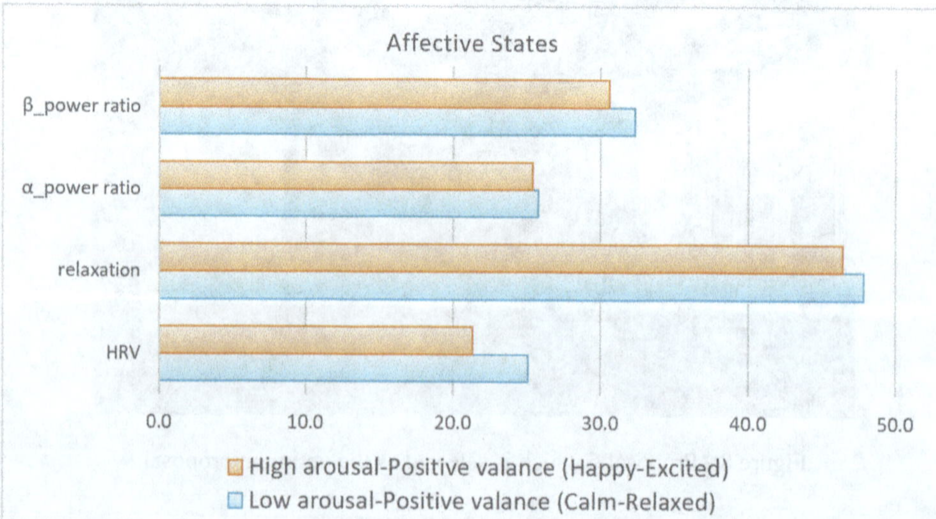

Figure 9: The SAM results and the output of the EEG headband.

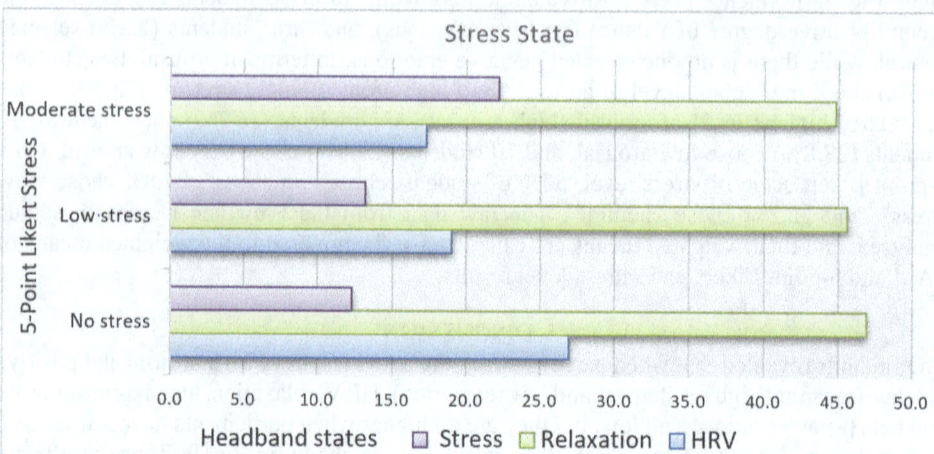

Figure 10: 5-point Likert stress results and the output of EEG headband.

While the methods used in this study are novel for the user-centered design approach, some limitations should be considered. One of the most important considerations in designing this study was the participant's affective mood state; they should not have a high level of depression in order for the results of EEG and HRV to be unaffected by depression; thus, PHQ-9 is required to decide whether to use mobile brain/heart imaging or not. Another issue is participants' willingness to wear the EEG headband without fear of being electrically shocked, as some students claimed; thus, the type of headband is critical to be simple and familiar in shape. This study aimed to use a consumer EEG headset that generates real-time affective states such as stress and relaxation so that searchers and architects could use it without the need for a programming team or complicated raw-data analysis of brainwaves. Future research should test other EEG headbands with more affective states, such as happiness, sadness, and so on, as well as simpler data analysis methods.

6 CONCLUSION

To summarize, the findings of this study show that a user-centered design approach is effective in built-environment design and leads to end-user satisfaction with the built environment. Furthermore, self-reporting measures such as questionnaires and a 5-point Likert scale can be used to determine end-user needs and evaluate design proposals. A 360-panorama displayed on a full HD laptop computer screen (non-immersive VR) is also the most appealing visualization technique for study participants. While end-users navigate non-immersive virtual reality, it affects their EEG and HRV and generates different affective states depending on the design. HRV, relaxation, and stress output of mobile brain/heart imaging while navigating 360-panorama have a direct relationship with a 5-point Likert scale of stress. Likewise, when participants in the SAM test have high arousal and positive valence (meaning high excitement and low relaxation), HRV, relaxation, alpha-power ratio, and beta-power ratio are all low, but they are all higher when participants have low arousal and positive valence (meaning high calm relaxation). Finally, using mobile brain/heart imaging to induce an end-user affective state while navigating design proposals in nonimmersive VR could improve design quality and end-user satisfaction.

ACKNOWLEDGEMENTS

All participants are students in the Department of Architecture at Alexandria University's Faculty of Engineering. The authors did all of the 3D modeling, VR creation, and data recording. This manuscript is not sponsored or endorsed by Flowtime, Inc., Lumion, Inc., or Autodesk, Inc.

REFERENCES

[1] Valand, M.S., Between organization and architecture: End-user participation in design. *Int. J. Work Organ. Emot.*, **4**(1), pp. 42–60, 2011.

[2] Farinango, C.D., Benavides, J.S., Cerón, J.D. & López, D.M. & Álvarez, R.E., Human-centered design of a personal health record system for metabolic syndrome management based on the ISO 9241-210: 2010 standard. *J. Multidiscip. Healthc.*, **11**, p. 21, 2018.

[3] El Sayad, Z.M., Farghaly, T. & Hamada, S.M., Integrating human-centered design methods in early design stage: using interactive architecture as a tool. *J. Al-Azhar Univ. Eng. Sect.*, **12**(44), pp. 947–960, 2017.

[4] Dwivedi, S.K.D., Upadhyay, S. & Tripathi, A.K., A working framework for the user-centered design approach and a survey of available methods. *Int. J. Scientific Res. Publ.*, **2**(4), pp. 12–19, 2012.

[5] Abisuga, A.O., Wang, C.C. & Sunindijo, R.Y., A holistic framework with user-centered facilities performance attributes for evaluating higher education buildings. *Facilities*, 2019.

[6] Vafaeenasab, M.R., Morowatisharifabad, M.A., Ghaneian, M.T., Hajhosseini, M. & Ehrampoush, M.H., Assessment of sick building syndrome and its associating factors among nurses in the educational hospitals of Shahid Sadoughi University of Medical Sciences, Yazd, Iran. *Glob. J. Health Sci.*, 7(2), p. 247, 2015.

[7] Simpeh, F., Comparative analysis of lecture theatre performance parameters' importance to students learning experience. *3rd Int. Conf. Infrastruct. Dev. Africa*, 2014.

[8] Duyan, F. & Ünver, R., A research on the effect of classroom wall colours on student's attention. *A| Z ITU J. Fac. Archit.*, **13**(2), pp. 73–78, 2016.

[9] Ahmadi, N. & Tabaeian, S.M., The role of color in improving the educational environment quality. Case Study: A number of girls' elementary schools in District 2 of Isfahan. *Int. J. Archit. Urban Dev.*, 7(2), pp. 33–40, 2017.

[10] Costa, M., Frumento, S., Nese, M. & Predieri, I., Interior color and psychological functioning in a university residence hall. *Front. Psychol.*, p. 1580, 2018.

[11] Sleegers, P.J.C., Moolenaar, N.M., Galetzka, M., Pruyn, A., Sarroukh, B.E. & Van der Zande, B., Lighting affects students' concentration positively: Findings from three Dutch studies. *Light. Res. Technol.*, **45**(2), pp. 159–175, 2013.

[12] Morrow, B.L., The impact of fluorescent and LED lighting on student attitudes and behavior in the classroom, 2018.

[13] Dadvand, P. et al., Green spaces and cognitive development in primary schoolchildren. *Proc. Natl. Acad. Sci.*, **112**(26), pp. 7937–7942, 2015.

[14] Determan, J. et al., The impact of biophilic learning spaces on student success. *Build. Res. Inf. Knowledgebase*, 2019. https://www.brikbase.org/content/impact-biophilic-learning-spaces-student-success.

[15] Bowman, C., Putney High School GDST The Biophilic Classroom Study – May 2020 –Final Report, 2020. DOI: 10.13140/RG.2.2.11556.50562.

[16] University of Toronto, Transforming the instructional landscape. 2019.

[17] Alonazi, M., Beloff, N. & White, M., Exploring determinants of m-government services: A study from the citizens' perspective in Saudi Arabia. *2019 Federated Conference on Computer Science and Information Systems (FedCSIS)*, pp. 627–631, 2019.

[18] Munoz-de-Escalona, E. & Canas, J.J., Online measuring of available resources. 2017.

[19] Horvat, M., Dobrinić, M., Novosel, M. & Jerčić, P., Assessing emotional responses induced in virtual reality using a consumer EEG headset: A preliminary report. *2018 41st International Convention on Information and Communication Technology, Electronics and Microelectronics (MIPRO)*, pp. 1006–1010, 2018.

[20] Suhaimi, N.S., Mountstephens, J. & Teo, J., EEG-based emotion recognition: A state-of-the-art review of current trends and opportunities. *Comput. Intell. Neurosci.*, 2020.

[21] Lakhan, P. et al., Consumer grade brain sensing for emotion recognition. *IEEE Sens. J.*, **19**(21), pp. 9896–9907, 2019.

[22] Wolverton, B.C., *How To Grow Fresh Air: 50 Houseplants To Purify Your Home Or Office*, Orion Spring, 2020.

[23] Fonseca, D., Villagrasa, S., Garcia, O., Navarro, I., Puig, J. & Paniagua, F., Statistical approach of user's experience in the visualization of architectural images in different environments. 2010.

[24] Joshi, A., Kale, S., Chandel, S. & Pal, D.K., Likert scale: Explored and explained. *Br. J. Appl. Sci. Technol.*, 7(4), p. 396, 2015.

ARE FLOATING CITIES THE URBAN FUTURE?

SAMRITHA YOGESH
Bartlett School of Architecture, University College London, UK

ABSTRACT

The vision of marine settlements evolved in the 19th century from the concept of living aboard a sailboat which is now known as sea steading. From thereon, the architecture community gave us a glimpse of innovative design experiments with strong statements which depict the agenda of revitalising the chaotic urban world. One of the most important and significantly urgent challenges for contemporary urbanisation is to develop their extent of sight beyond the political and social insights of the city to address changing morphological terrains of urbanization. The transformation of the world that the 19th century created is the framework for the global systems which are continued even today. The futuristic vision of the utopian architects in the 1950s gave a path to believe that urban settlements could be built in extreme geographical conditions like the polar regions, the deserts and on the sea. The floating cities are still not a reality, but the longstanding vision is gradually making its way to become the urban future in the hope to sort the contemporary urban issues starting from the scarcity of land at global level due to over population to the drastic climate change due to the exploitation of nature by humans. Floating cities clearly intend to become a crucial architectural element in the future urbanization. The theory behind the concept of future urbanization is intended to help explore the unexplored morphological terrains for urban restructuring around the world. In this paper floating cities will be analysed to understand as to how "self-sufficient floating cities" could combat contemporary urban issues and also influence future urbanisation.

Keywords: floating cites, urban future, urban environment, amphibious architecture.

1 INTRODUCTION

Could "self-sufficient floating cities" combat contemporary urban issues?

The city is the necessary instrument for the evolution of humankind [1]. But due to the rapid urbanisation the urban cities have been deteriorating mainly due to human activities which affects the environment and then eventually the humankind. The main urban issues faced are increasing population and rising sea levels due to drastic climate change which also has a major impact on the earth's ecosystem. 60% of the global urban population will be residing in the cities by 2030 [2]. This is going to be very stressful for the urban cities as they must battle with the population against extreme climate changes. 90% metropolitan cities are situated in the coastal regions which are vulnerable to the rising sea level caused by greenhouse gases leading to the sinking of the cities by 2100 [2]. Even if we reduce the greenhouse gas emission which will be very difficult to do so because of the world population which estimates to about 8 billion people and very soon it will become 10 billion people, sea levels are still going to rise [3]. Urban cities and their spaces play a very important role in planetary urbanisation with politics, economy, social life, culture and environmental conditions being the influencing factors [4].

2 FLOODING

The major impact of global warming is sea level rise which will have adverse effects on the coastal cities leading to its disappearance (Fig. 1). This has been a major environmental concern because from 1900s the flooding has been increasing gradually with impact being severe between 1981 to 2010 and its from here its gradually increasing till date. In 2010 approximately 178 million people were affected by floods and the economy losses incurred

Figure 1: Impact of flooding.

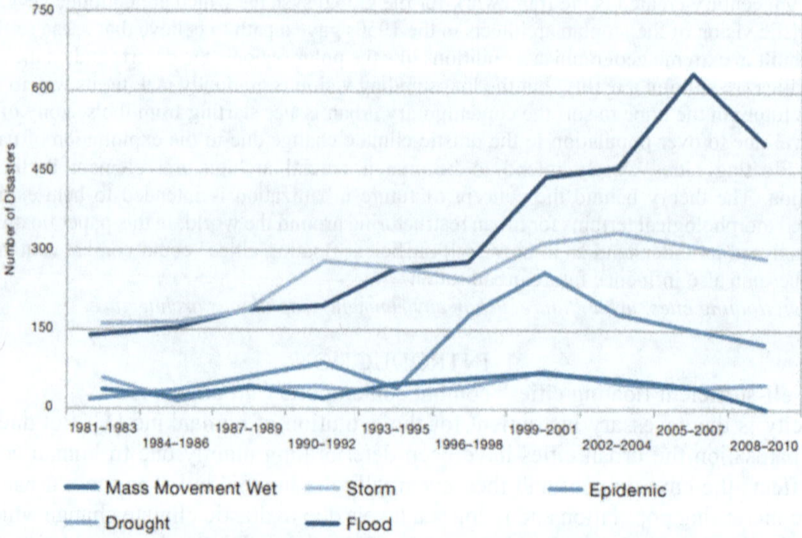

Figure 2: Global data of flooding and its effects between 1981–2010.

between 1981 to 2010 was over 40 billion dollars [5] (Fig. 2). The year 2021 the economy losses incurred was over 112-billion-dollar due to floods which means the flooding and the economy loss is increasing thrice the rate of the previous decade.

3 FLOOD RISK MANAGEMENT

Flooding is a disaster which affects urban environment in a huge way. Over the years a lot of research has been done to tackle flooding. From the early 2000, number of design methodologies have been surfacing in the context of tackling floods. The main idea behind the innovation of these methodologies is that flooding is inevitable because of the sea level rise so adapting is the best way to approach the disaster so thereby instead of looking at it as a disadvantage its better we find ways to prevent and adapt to flooding as well plug it in the urban environment in such a way that it becomes part of the urban setting.

Below are mentioned few design methodologies that tackle flooding.

3.1 Sponge cities

Sponge cites are a type of cities which are designed mainly to tackle flooding issues. The interesting part about this design methodology is that the disaster is very swiftly made into an advantage which ultimately contributes to the urban environment. Ideally floods cause damage but in this case when it floods the water is absorbed and then recycled so that it can be used for urban activities. When the city gets flooded it acts like sponge and absorbs all the water which are stored in urban aquifers before which it is filtered through the soil [6] (Fig. 3). This methodology is a practical solution for flooding.

Figure 3: The Wujiang river in Zhejiang, Yu's home province.

3.2 Floating eco-homes

Floating Eco-Homes-Harnaschpolder is a project a residential building built in the water retention area of delft where the main objective was to construct homes on water by using energy efficient systems and sustainable design approach to reduce the impact on the environment (Fig. 4). A distinctive design feature of this project was that the citizens own a plot on water. These successful small-scale projects gave the architects and urban designers a boost to use the developing technology to their advantage to design self-sufficient floating cities which propagates off the grid habitation. The water quality is checked by the water measurement program using underwater drones. This experimental project was built in a flood prone area, where only 6 houses were built to check whether this design methodology worked, and it turned out to be successful because it tackled flooding and the residents were very comfortable as they didn't feel any stress staying in the floating home.

3.3 Floating Pavilion, Rotterdam

The "Floating Pavilion Rotterdam" was one of the most innovative projects done by delta sync which has become an international exemplar for its sensitive eco-friendly design towards the environment and for using sustainable design ideologies [7] (Fig. 5). It's an interactive conference and exhibition space which is designed in such a way that it acts like a bridge between city which is land and the port which is water. The pavilion is energy efficient; it gets heated and cooled down naturally without any artificial interference. The pavilion reacts with climate in such a way that it gets heated by solar energy and it cools

Figure 4: Floating Eco-Homes-Harnaschpolder.

Figure 5: Floating Pavilion, Rotterdam.

down with the surface water. This project was an experiment to analyse if floating buildings can be viable in an urban environment and can they replace buildings which are sinking due to the sea level rise.

3.4 Amphibious architecture

It's a design methodology which is used to tackle flooding. The building is designed in such a way that when the area gets flooded it will rise to water level with the help of a hydraulic system and then it will act like a floating building. This structure is built on four vertical posts with the floating blocks placed beneath the house, when flooding occurs the floating blocks lift the building hydraulically. This has been successfully implemented in many floods prone area starting from New Orleans [8] (Fig. 6).

Figure 6: Amphibian building, Orleans.

3.5 Floating cities

Floating cities has been in the limelight for a very long time as proposals, but none have been implemented, projects which have been implemented are small scale residences or pavilions which has become a strong foundation for the implementation of floating cites. These successful small-scale projects gave the architects and urban designers a boost to use the developing technology to their advantage to design self-sufficient floating cities which propagates off the grid habitation (Fig. 7). The aim behind the design of floating cities will be reconfiguring cities with innovative urban spaces and implementation technology innovations [9]. Floating cities can be a replacement for the cites that are sinking due to flooding.

Figure 7: Oceanix City – Floating city proposal by BIG (Bjarke Ingels Group).

4 WHY FLOATING CITIES?

Studies say if 13% of the human population start to live on water, humanity will achieve harmony with nature which requires 1% of the total ocean area. "Floating cities could be repeatedly remoulded according to the seasons or population changes" [2]. This enables flexibility in the expansion of the city with the urban development. Floating cities gives an opportunity to free the urban cities from the housing crunch by providing prefabricated low-cost housing on the sea which would create a cost-efficient housing [10]. Urban designers saw a potential to re-establish urbanisation in a more sustainable approach through the emergence of the floating cities as they would be designed on completely new sustainable principles. Self-efficient floating cities can also be the solution to the water–food–energy nexus and give an opportunity to create nations which will govern independently on the sea giving rise to ocean communities and ecosystem made by mankind that are designed to develop, evolve, and change with time and it would be termed as Seavilisation.

5 ARCHITECTURAL EVOLUTIONS OF FLOATING CITIES

5.1 Beginning of floating cities

The idea of Floating cities and design development evolved from the already existing floating villages like the Tonle Sap, Cambodia, Ko Panyi, Thailand, Uros, Peru, Day Asan, Philippines, Ha Long Bay, Vietnam and Ganvie, West Africa (Figs 8 and 9). These floating villages has been developed keeping its geography and climatic conditions in mind, but beyond the morphological conditions the social and cultural aspects has also influenced the development of the villages.

Figure 8: Ko Panyi, Thailand.

Figure 9: Ganvie, West Africa.

5.2 Influence of utopian architecture

The architectural projects done in the 1950s by utopian architects like Tokyo Bay Plan designed by Kenzo Tange in 1960 also called the "The Japanese Metabolists", Buckminster Fuller's Triton City and Archigram and Paolo Soleri which proposed underwater sea farms were few among the utopian projects which gave a platform for the current architects and engineers to think about making a floating city a reality (Figs 10 and 11). These projects worked around the fluidity of the sea which gave rise to floating cities designed in an organic way.

Figure 10: The Japanese Metabolists.

Figure 11: Triton City.

5.3 Emergence of floating cities

In the early 2000s, when climate change was seriously impacting the world and a lot of coastal cities facing its consequences from here came the urgent need to come up with solutions for the sinking cities. Few of the design solutions were that the building retention systems, displacing the coastal settlement or rising the settlement [11]. On small scales amphibious architecture and floating architecture was implemented. Using amphibian concept, the buildings were designed as such that it rises with the flooding and goes down when the water is drained out, this style was useful in zones where water gets drained after some time [9] (Fig. 12). The other one was floating architecture where the buildings are designed in such way that it permanently floats on the water, and this works in zones were flooding of the settlement is permanent (Fig. 13). Keeping the above aspects and implemented design methodologies the floating cities proposal emerged through merging, modifying and reassembling of different concepts with the available support of technology.

Figure 12: Waterloo, Ontario – NRC research pavilion (amphibian architecture).

Figure 13: Scoonship – A floating settlement built in Amsterdam.

6 SPECULATIONS ABOUT THE FLOATING CITES

Floating cites were speculated to become a market of governance. Instead of following the footsteps of global socialism, the floating cities will open a market of societies with different ideologies, culture, governing styles and political outlooks. Different societies can compete for citizens [8].

7 SOVEREIGN FLOATING CITIES AND MICRONATIONS

While designing a floating city every aspect has been taken into consideration but there is one aspect which doesn't come a long way in the vision which is the social political boundaries. The proposals don't specify under which governing authority or global government the city would come under which means it gives an opportunity for the city to build its own government with legal regulations formed based on the urban design, culture, lifestyle, economy and the local environment. So, the floating cites can have functional government body with social structures that are independent from external interference which will ideally be called a micronation. There was an experiment done called Principality of Sealand in the North Sea as it was claimed as an independent offshore micronation since 1967 [12] (Fig. 14). The platform is a military fortress called rough towers which was under British government during world war but now it's owned by the Bates family. This micronation has its own centralised government which is free from the external interference. The micronations can form the structure of the government in relation to the traditions of anarchism and libertarianism [13]. Instead of having districts, boroughs or states the floating cites will divided into ecological enclaves [14].

8 THE CHALLENGES IN BUILDING FLOATING CITIES

The main challenges floating cities would face in the implementation will be site selection and cost of construction which is turning out to be very expensive because of the state-of-the-art technology, sustainable materials and constructing the city as disaster resistant module which is very expensive. Since the population belonging to all the different economical categories in the cities must gradually settle in the floating city so thereby the housing should

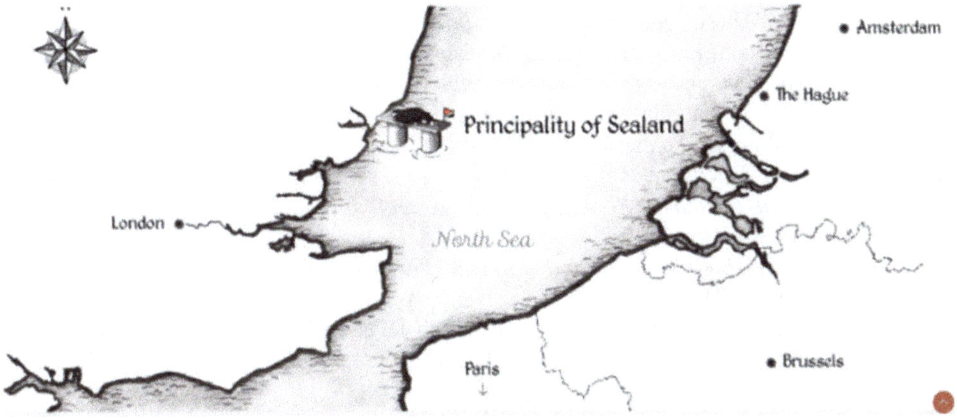

Figure 14: Principality of Sealand.

be affordable, but the cost of construction is becoming an obstacle and eventually there is a possibility that the floating cities become an expensive development which is only exclusive to the elites with power, resources and money.

9 CONCLUSIONS

The implementation of floating cities will mark a new beginning for the future of urbanisation. Floating cites can freely actualize in a unified manner to maintain world order [15]. Floating cites as the power to change the geography of the world with new nations added. But behind this design proposal there is an aspect which is not addressed which is the impact on the marine ecosystem which is the most sensitive component of the planet. Covering ocean surface would bring risk to the marine ocean system because light plays an important role in the natural biochemical cycles and food chain. The latest proposal for floating cities does propagate eco-friendly design approach but the impact on the ocean cannot be ruled out 100%, thereby the impact as to be studied in detail to find adequate solutions, but this can be achieved only when it's implemented because nothing substantial will come out based on speculations. As time changes, we will gradually understand the impact of the floating city on the urban and environmental condition and analyse whether the path taken to design and implement the city is an eco-friendly approach but for now the floating city proposes to be the urban future that will be an exemplar for the other developing cities also a solution to urban issues.

ACKNOWLEDGEMENTS

I would like to thank my history and theory tutor, Mr. Huang William for his dedicated support, guidance and encouragement throughout my work. I would also like to thank my parents for their unconditional support and encouragement which helped me complete my work.

REFERENCES

[1] Die, M.O., Miniaturize or Die! Paolo Soleri's City as Architecture.
[2] Cosgrave, E., The future of floating cities–and the realities. *BBC Future,* 11, 2017.
[3] Englander, J., *Moving to Higher Ground: Rising Sea Level and the Path Forward*, The Science Bookshelf: Boca Raton, 2021.

[4] Brenner, N., Theses on urbanization. *Public Culture,* **25**, pp. 85–114, 2013.
[5] Jha, A.K., Bloch, R. & Lamond, J., *Cities and Flooding: A Guide to Integrated Urban Flood Risk Management for the 21st Century*, World Bank Publications, 2012.
[6] Garfield, L., China is building 30 "sponge cities" that aim to soak up floodwater and prevent disaster. *Business Insider*, 2017
[7] Quirk, J. & Friedman, P., *Seasteading: How Floating Nations Will Restore the Environment, Enrich the Poor, Cure the Sick, and Liberate Humanity from Politicians*, Simon and Schuster, 2017.
[8] Gunn, N., *Sinking Cities: Adaptive Architecture for Rising Sea Levels,* University of Washington, 2021.
[9] Abbott, K.W. & Bernstein, S., The high-level political forum on sustainable development: Orchestration by default and design. *Global Policy,* **6**, pp. 222–233, 2015.
[10] Revkin, A., Floating cities could ease the world's housing crunch, the UN says, 2019.
[11] Hill, K. & Barnett, J., Design for rising sea levels. *Harvard Design Magazine,* **27**, pp. 1–7, 2007.
[12] Strauss, E.S., *How to Start Your Own Country: How You Can Profit from the Coming Decline of the Nation State*, Loompanics Unlimited, 1984.
[13] White, D.F. & Kossoff, G., Anarchism, libertarianism and environmentalism: Anti-authoritarian thought and the search for self-organizing societies. *The SAGE Handbook of Environment and Society*, pp. 50–65, 2007.
[14] Hodson, M. & Marvin, S., Urbanism in the anthropocene: Ecological urbanism or premium ecological enclaves? *City,* **14**, pp. 298–313, 2010.
[15] Steinberg, P.E., Nyman, E. & Caraccioli, M.J., Atlas swam: Freedom, capital, and floating sovereignties in the seasteading vision. *Antipode,* **44**, pp. 1532–1550, 2012.

USE OF ENGINEERED WOOD FOR THE RETROFITTING OF EXISTING STRUCTURES

GIORGIO FRUNZIO[1]*, SERGIO RINALDI[1]†, MARIATERESA GUADAGNUOLO[1]‡, LUIGI MASSARO[1]§ & LUCIANA DI GENNARO[2]**
[1]Department of Architecture and Industrial Design, University of Campania Luigi Vanvitelli, Italy
[2]Department of Engineering, University of Campania Luigi Vanvitelli, Italy

ABSTRACT

Densely populated cities characterize urbanized territories, and most public buildings fail to meet many of the new demands imposed by the pandemic and the modern lifestyle. The need to expand spaces has become imperative, although, at the same time, it is necessary to reduce land consumption and preserve green spaces. If we consider the existing heritage a resource, these problems could represent an opportunity to improve existing structures both from a technological and structural point of view. It is possible to effort a holistic approach by improving the energy impact of new buildings and the heritage seismic behaviour facing the problem in a multidisciplinary way. Starting from the "building on the built" philosophy, this paper presents a possible use of engineered wood, such as cross-laminated timber (CLT), to pursue this strategy. The proposal is the realization of volumetric additions to existing buildings without further land consumption. This type of intervention, mentioned as "parasitic architecture", positively impacts urban regeneration strategies. The use of prefabricated timber components (CLT) endorses the speed of realization, reducing the interferences with the surroundings and improving the healthiness and safety of the environment.

Keywords: parasite architecture, sustainable architecture, structural analysis, cross laminated timber, CLT, engineered wood, pres-lam, life cycle assessment, LCA, footprint.

1 INTRODUCTION

Today the need to expand anthropized spaces is a reality with which is necessary to interact. At the same time, the reduction of soil consumption appears fundamental to preserve green spaces. Within the scenery of possible solutions to the problem, the intervention on existing buildings can represents an opportunity. It could be an occasion both to adequate the seismic response of existing buildings according to current rules (NTC 2018) both to improve their energy impact [1]. The presence of volumetric additions to existing buildings could play the role of renewing the formal, structural, and technological aspects. The loads increasing on the structures requires a deep knowledge of the buildings to better evaluate the new static condition moreover [2]. Even if, for existing building, local reinforcements represent the better solution [3]. It is possible to effort a holistic approach by:

- improving the energy impact of new buildings;
- improving the heritage seismic behaviour;
- saving green space;
- reducing consumption of land;
- facing the problems in a multidisciplinary way [4].

* *ORCID: http://orcid.org/0000-0001-7175-0608*
† *ORCID: http://orcid.org/0000-0002-4469-4601*
‡ *ORCID: http://orcid.org/0000-0002-8273-4846*
§ *ORCID: http://orcid.org/0000-0003-2755-5471*
** *ORCID: http://orcid.org/0000-0002-6435-7290*

WIT Transactions on The Built Environment, Vol 210, © 2022 WIT Press
www.witpress.com, ISSN 1743-3509 (on-line)
doi:10.2495/ARC220191

Starting from the "building on the built" philosophy, there are three types of intervention:

- Parasitical architecture (Fig. 1). New volumes are added onto existing buildings but distinct from the host one both formally and spatially but linked to.
- Rooftop architecture (Fig. 2). Roofs are used as new soils to create a local expansion or to design a complex urban system [5].
- Exoskeleton systems. It can be superimposed on the facades by creating an independent structure on its foundations. It can accommodate new spaces and support possible elevations [6].

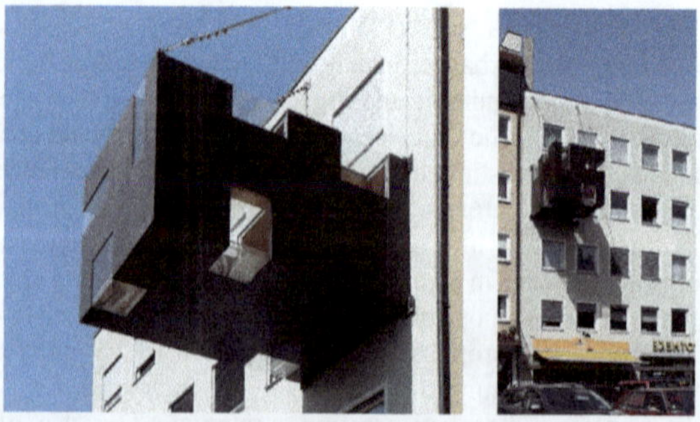

Figure 1: Two view of Rucksack House arch. Stefan Eberstadt, 2004.

Figure 2: Enlargement of the University of Graz, Atelier Thomas Pucher, 2019.

The proposal of this paper is to investigate all the aspects involved the volumetric additions to existing buildings providing a possible solution to expansion without further land consumption. The most important aspect, according to our point of view, is the challenge of the building material choice which deeply influences both structural and functional aspects of the intervention [3], [7]. From this perspective, the use of wood or its technological evolutions as engineered wood, could be a real way to pursue the holistic approach. The main features of the engineered wood are:

- respect high structural and architectural standards;
- increase energy efficiency;
- increase environmental sustainability;

- speed of installation;
- dry processing.

A brief overview of the authors proposal, is presented in the following section.

2 PARASITE ARCHITECTURE TECHNOLOGIES

"Parasitical Architecture" can represent "building on the built" philosophy for a sustainable urban regeneration. The new added volumes can be realized according to the following schemes (Fig. 3).

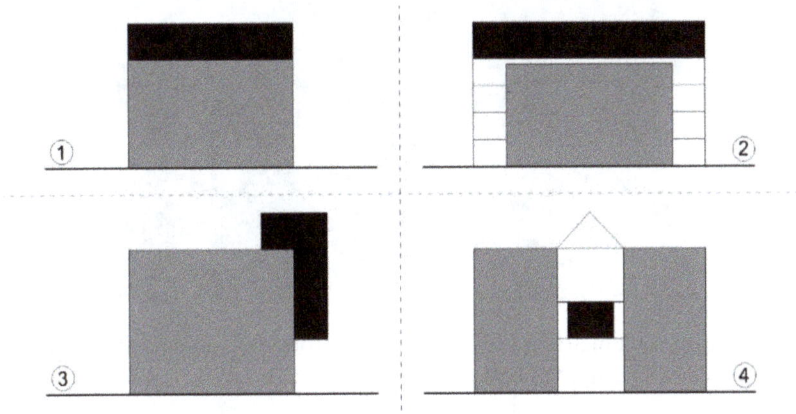

Figure 3: Schemes of parasitic volumes.

The first three schemes represent additions to the existing building while the last one provides a new volume within the space that elapses between two separate structures.. So, without consumption of land, each one creates social and common areas.

In the first scheme the new volume is placed on the roof top and is entirely supported by the existing structures. It is realized with the same characteristics and technologies of the underlying building reproducing its configuration without variations of materials and shape except for the extra height.

In the second scheme, the new volume often named exoskeleton, is completely separated from the original building with independent foundation (Fig. 4). It a self-supporting structure and any load increase is not transferred to the main building.

Figure 4: Addition with exoskeleton for A. Volta Institute of Aversa (CE) Italy: Actual state and rendering of the project state [8].

The third scheme best fits the idea of "parasite architecture". It is an addition that can be supported by the facades, the roof tops or both (Figs 5 and 6).

Figure 5: Parasitic volumetric additions in the district "le fornaci" in S. Maria Capua Vetere (CE) Italy: State of fact [9].

Figure 6: Parasitic volumetric additions in the district "le fornaci" in S. Maria Capua Vetere (CE) Italy: Rendering of the project state [9].

They are usually designed as expansions of housing units or to create new common independent areas vertical connected with the original structure.

The last scheme is an "aerial" volume located in the free courts. It provides aero-illuminating ratios without using free areas usable for outdoor activities. It could be a good solution for school buildings with large courtyards. With this solution, schools earn new areas keeping the spaces for outdoor activities unchanged.

Large span beams connected to the internal perimeter walls support the new volumes giving the structural elements of the walkway the role of horizontal links.

The main difficulties that this typological approach could produce are synthetically listed as follow:

- the connection between the new volumes and the existing one;
- the architectural, environmental, social impact of the new volumes on the urban context;
- the evaluation of the real improvement in terms of accessibility, energy efficiency, safety, new services, etc.;
- the addition of new loads on the load-bearing structure could create problems to its foundation system.

It has to be highlighted that if the basic structure is constituted of masonry, some further consideration must be made especially for the scheme 3 and 4. This problem is better exposed in the following paragraph.

3 PARASITE ARCHITECTURE: STRUCTURAL ASPECTS

It is not possible to have complete knowledge of the existing structure only through the determinations of all mechanical parameters that manage it, especially for masonry buildings [10]–[13]. These parameters should be considered together with other information like technologies used at the construction time, structural schemes, and design choices. Any volumetric addition imposes an interaction between new and old. As shown, the new addition can be connected or not to the original building. So, the possible scheme depends on the type of addendum. Schemes 1 and 3 best represent the parasitic dependence of the "host" structure. The new configuration of the building, according to the Italian standard NTC 2018 [14], is subject to global verification mainly due to the load increase on the original structure [15], [16]. The increase in masses, the "new" centre of gravity and of stiffnesses and the "new" shape ratios impose a modification of the seismic behaviour, however. So, the increasing of load-bearing sections for scheme 1 and the eccentric added mass in scheme 3 determine a reasonable modification of the static schemes. The last case requires in-depth assessments of the torsional behaviour of the building, moreover [17], [18]. An example of eccentric volumetric addition, or simply scheme 3, is shown in Figs 7 and 8. This new volume is built off-site and added by tie rods and bolts to the R/C base structure. If the base structure is a masonry one, it must be reinforced at the contact points [19], [20]. In any case, a local verification is required for the anchors. Fig. 9 shows the two possible solutions:

- on the two facades of a linear building;
- on the facades of an internal courtyard.

Scheme 2 (exoskeleton) is an independent structure from the existing one. Even if the new addition could share the foundation of the old structure. It is also important to consider the soil–structure interaction (SSI) [21], [22]. This case has similar structural problems of schemes 1 and 3, even if the addendum does not completely depend on the base structure. This problem could represent a possibility to reinforce the existing building (Fig. 9) by acting as an active element in the structural retrofit strategy [23]. The adaptive exoskeleton system [24] is an example of this intervention strategy (Fig. 10). It can be viewed as a device that improves the structure's initial characteristics.

When the addendum is completely independent from the original building, the real challenge is keeping independence unchanged over time. The distance between the

Figure 7: Render of a parasitic volumetric additions of type 3.

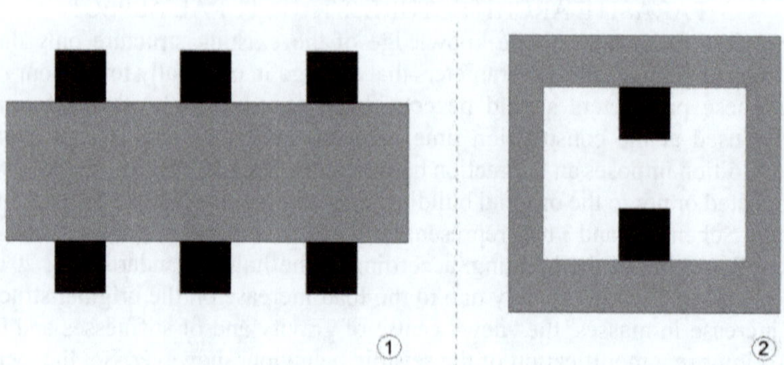

Figure 8: Parasitic volumetric additions of scheme 3: possible layout in plan (1) linear building, (2) courtyard building.

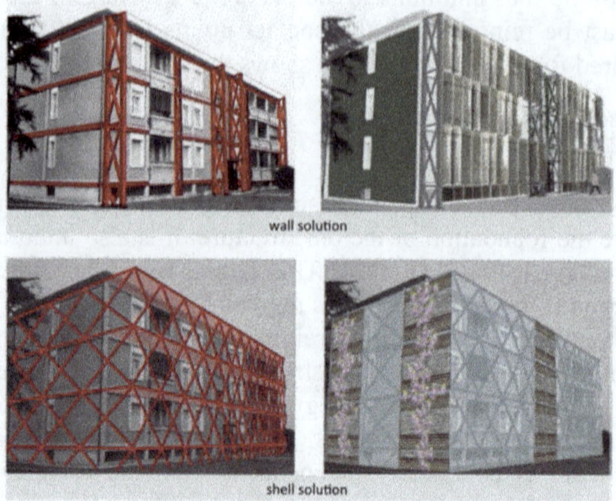

Figure 9: Two retrofit solutions with adaptive exoskeleton [23].

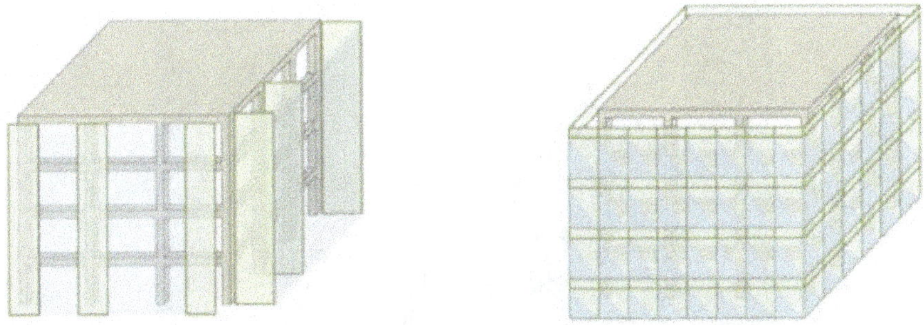

Figure 10: Two retrofit solutions with added seismic-resistant elements [24].

foundations of the two structures must be adequate to avoid that the structures interact during an earthquake. The design of the seismic joints is a mandatory [25]. The construction of suitable structures to climb over the original one meets some difficulties. The new structure has to guarantee compliance with the requirements and their maintenance over time.

Large decks impose the improvement of stiffness of the floors. Adequate qualities of translating bi-directional stiffness and resistance as well as torsional stiffness qualities could be achieved by the best plan position of concentric bracing [26].

One other problem is represented by pillars that weigh on the beams of the over structures. So, the “floating column” behaviour imposes an in-depth analysis that considers:

- the effects of the vertical component of the seismic action;
- the punching problem;
- the torsional effects additional.

Villa Fiorita Clinic located in Aversa [27] is an example of addendum with an independent structure (Figs 11 and 12). There is no interaction between the new and old which is completely covered and incorporated by the new. In case of earthquake the interactions between the buildings is reduced by dissipation devices which allows to limit the displacement of the addendum. The concentric bracing with active diagonals allows to have an earthquake-resistant structure and the perform the dissipative function (Figs 13 and 14).

Figure 11: Villa Fiorita Clinic: State of fact [27].

Figure 12: Villa Fiorita Clinic: Project state [27].

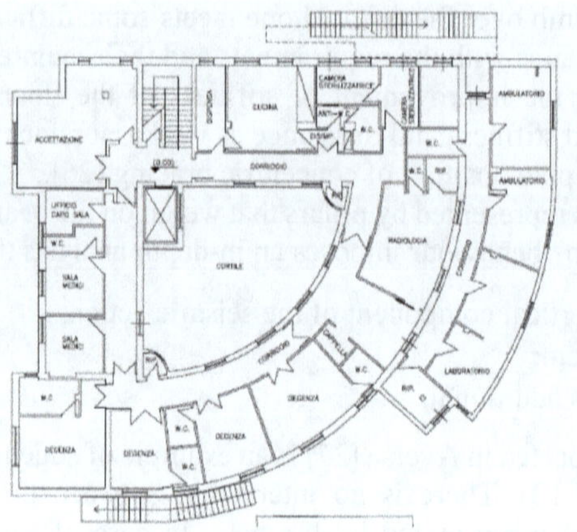

Figure 13: Villa Fiorita Clinic: Original plan view [27].

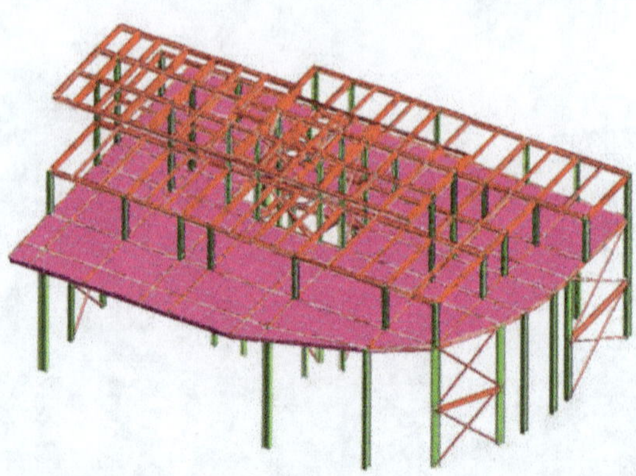

Figure 14: Villa Fiorita Clinic: Structural model of the added structure [27].

Scheme 4, the last one, is the most complex scheme due to the difficulties to establish the new connections to the existing structure. Moreover, the structural elements, which impose horizontal actions, need major attention.

This concise overview reveals that schemes 2 and 4 can be used to improve the structural performances and to increase the seismic capacity of the building. To significantly reduce post-event repair costs [28], or simply to apply the "low-damage" strategies, the schemes proposed use dissipative elements and systems which let obtain the increasing of the seismic capacity and limit post-event damage. A way to achieve the goals proposed is the use of wood [29], [30]. The problems related to the anisotropy of raw material has been overcame by technological innovations that give birth to the engineered wood that can fully be used as a valid alternative of the most commonly structural materials [11], [31]–[33]. The use of wood could create smart and sustainable buildings with the volumetric additions without soil consumption.

Another possible solution is the pres-lam system [34] created for seismic-resilient and sustainable structures [35]. This patented technology is based on the use of, glulam and/or CLT (cross-laminated timber) that are prestressing by a post-tensioning system on-site using sliding cables and tensioning equipment.

The dissipative characteristics of these structures made with this technology are investigated by several authors [36]–[38]. Also, the dissipative effects due to the insertion of rocking elements inside existing structures [39]–[43], or the presence of metal connections for reinforcement interventions in its plane with CLT panels [44] or off-plan [45] are investigated. There is a common trend in the in field research groups that investigate the possible use of this elements for exoskeletons or endoskeletons on existing masonry buildings [46], [47].

The use of engineered wood, like, CLT panels, allows a holistic approach for the Eco Architectural Renovation, not only in energy and seismic terms but also as a low impact during the construction phases. The use of these elements, substantially executed off-site, allows:

- faster installation;
- the precisely establishment of the times;
- the clear identification of the areas involved in the construction and assembly phases;
- above all, the original building could keep its activities without interruption.

Moreover, the use of wood let to strongly reduce the production of CO_2, using wood as building material imposes the management of forest that encourage the growth of new trees. The wood used to realize the CLT panels keep storing di CO_2. To this amount we can add the amount saved choosing wood rather than another building material. The life cycle assessment related to wood allows to achieve low CO_2 footprint intervention according to eco-architecture and same goals proposed in agenda 2030.

4 CONCLUSIONS

The use of engineered wood like CLT and pres-lam, assembled with innovative "off-site" technologies, let the realization of volumetric addition without further free soil consumption. Moreover, prefabricated elements allow rapid interventions even in densely populated urban areas. This reduces the interference with the surrounding as a safety advantage. The parasite architecture concept fits the "resilient thinking" transforming the need to expand the spaces as an opportunity to improve the energy impact and structural performance of the existing heritage, with a great multidisciplinary approach and with low CO_2 footprint. Two ways are identified:

- the advantage of a lower regulatory burden adopting the added independent of the base building;
- the improvement of seismic characteristics of the base building where additions become one with it.

REFERENCES

[1] Arena, M., Cannaviello, M., Rinaldi, S. & Violano, A., Retrofit vs rehabilitation: Improving technological quality of the existing school building. La Scuola di Pitagora, Napoli, 2011.

[2] Monaco, M., Iannuzzo, A., Tafuro, A. & Gesualdo, A., Dynamic analysis of a Pompeian Domus. *XI International Conference on Structural Dynamics*, pp. 4922–4929, 2020. DOI: 10.47964/1120.9399.19876.

[3] Bergamasco, I., Gesualdo, A., Iannuzzo, A. & Monaco, M., An integrated approach to the conservation of the roofing structures in the Pompeian Domus. *Journal of Cultural Heritage*, **31**, pp. 141–151, 2018. DOI: 10.1016/J.CULHER.2017.12.006.

[4] Frunzio, G., Guadagnuolo, M., Massaro, L. & di Gennaro, L., The CLT panels: A sustainable response for existing buildings. *Proc. Beyond All Limits. International Conference on Sustainability in Architecture, Planning, and Design*, 11–13 May, Belvedere Di San Leucio, Caserta, Italy, 2022.

[5] Khalfi, O., Rooftop architecture: A sustainable alternative for urban expansion. Temple University, 2019.

[6] di Giulio, R., Sustainable roof-top extension: A pilot project in Florence. *Portugal SB10 Sustainable Building Affordable to All*, pp. 255–264, 2010.

[7] Violano, A. et al., Teaching technological design: Enhancing strategies and approach. *12th International Conference on Education and New Learning Technologies*, pp. 8094–8103, 2020. DOI: 10.21125/edulearn.2020.2024.

[8] Fabozzi, R., Addizione con esoscheletro per l'Istituto A. Volta di Aversa (CE) Italia. Università per gli studi della Campania Luigi Vanvitelli.

[9] Borriello, G.E., Addizioni volumetriche parassite nel quartiere "le fornaci" a S. Maria Capua Vetere (CE) Italia, 2015.

[10] Frunzio, G., di Gennaro, L. & Guadagnuolo, M., Palazzo Ducale in Parete: Remarks on code provisions. *International Journal of Masonry Research and Innovation*, **4**, p. 159, 2019. DOI: 10.1504/IJMRI.2019.096826.

[11] di Gennaro, L. & Frunzio, G., Wood in the structural restoration of masonry buildings. *Le Vie Dei Mercanti XVII International Forum – World Heritage and Legacy*, pp. 934–942, 2019.

[12] Guadagnuolo, M., Aurilio, M., Basile, A. & Faella, G., Modulus of elasticity and compressive strength of tuff masonry: Results of a wide set of flat-jack tests. *Buildings*, **10**, 2020. DOI: 10.3390/BUILDINGS10050084.

[13] Monaco, M., Aurilio, M., Tafuro, A. & Guadagnuolo, M., Sustainable mortars for application in the cultural heritage field. *Materials*, **14**(3), p. 598. 2021. DOI: 10.3390/ma14030598.

[14] NTC 2018, Aggiornamento delle Norme tecniche perle costruzioni. 2018.

[15] Guadagnuolo, M., Aurilio, M. & Faella, G., Retrofit assessment of masonry buildings through simplified structural analysis. *Frattura Ed Integrità Strutturale*, **14**, pp. 398–409, 2020. DOI: 10.3221/IGF-ESIS.51.29.

[16] de Angelis, A., Tariello, F., de Masi, R.F. & Pecce, M.R., Comparison of different solutions for a seismic and energy retrofit of an auditorium. *Sustainability*, **13**(16), p. 8761, 2021. DOI: 10.3390/su13168761.

[17] de Stefano, M. & Mariani, V., Pushover analysis for plan irregular building structures. *Geotechnical, Geological and Earthquake Engineering*, **34**, pp. 429–448, 2014. DOI: 10.1007/978-3-319-07118-3_13/FIGURES/6.

[18] Faella, G., Giordano, A. & Guadagnuolo, M., Unsymmetric-plan masonry buildings: Pushover vs nonlinear dynamic analysis. *Proceedings of the 9th U.S. National and 10th Canadian Conference on Earthquake Engineering*, 25–29 Jul, Toronto, Canada, 863, 2010.

[19] Calderoni, B., Prota, A., Cordasco, E.A. & Sandoli, A., Seismic vulnerability of "ancient" masonry buildings and strengthening intervention strategies. *Proceedings of the 16th International Brick and Block Masonry Conference*, 26–30 Jun., Padova, Italy, pp. 727–736, 2016. DOI: 10.1201/b21889-99.

[20] Guadagnuolo, M. & Faella, G., Simplified design of masonry ring-beams reinforced by flax fibers for existing buildings retrofitting. *Buildings*, **10**, p. 12, 2020. DOI: 10.3390/buildings10010012.

[21] de Angelis, A., Mucciacciaro, M., Pecce, M.R. & Sica, S., Influence of SSI on the stiffness of bridge systems founded on caissons. *Journal of Bridge Engineering*, **22**, 04017045, 2017. DOI: 10.1061/(ASCE)BE.1943-5592.0001073.

[22] de Angelis, A., Lourenço, P.B., Sica, S. & Pecce, M.R., Influence of the ground on the structural identification of a bell-tower by ambient vibration testing. *Soil Dynamics and Earthquake Engineering*, **155**, 2022. DOI: 10.1016/J.SOILDYN.2021.107102.

[23] Marini, A. et al., Combining seismic retrofit with energy refurbishment for the sustainable renovation of RC buildings: A proof of concept. *European Journal of Environmental and Civil Engineering*, **26**, pp. 2475–2495, 2022. DOI: 10.1080/19648189.2017.1363665.

[24] Bellini, O.E., Adaptive exoskeleton systems: *Remodelage* for social housing on Piazzale Visconti (BG). *Regeneration of the Built Environment from a Circular Economy Perspective*, eds S. Della Torre, S. Cattaneo, C. Lenzi & A. Zanelli, Springer: Cham, pp. 363–374, 2020. DOI: 10.1007/978-3-030-33256-3_34.

[25] Frettoloso, C. & Guadagnuolo, M., Seismic joints: Architectural integration and structural safety. *XIII International Forum Le vie dei Mercanti – Heritage and Technology – Mind Knowledge Experience*, Scuola Pitagora Editore, Napoli, 2015.

[26] Alecci, V., de Stefano, M., Galassi, S., Lapi, M. & Orlando, M., Evaluation of the American approach for detecting plan irregularity. *Advances in Civil Engineering*, pp. 1–10, 2019. DOI: 10.1155/2019/2861093.

[27] Frunzio, G., Di Rosa, L., Lanzotti, C. & Lampitiello, S., Sopraelevare senza toccare tower raised without affecting. *XXIII Giornate Italiane Della Costruzione In Acciaio – Realizzazioni*, pp. 139–146, 2011.

[28] Buonocore, G., Gesualdo, A., Monaco, M. & Savino, M.T., Improvement of seismic performance of unreinforced masonry buildings using steel frames. *Proceedings of the 12th International Conference on Computational Structures Technology*, 117, 2014. DOI: 10.4203/ccp.106.117.

[29] Valluzzi, M.R. et al., Nested buildings: An innovative strategy for the integrated seismic and energy retrofit of existing masonry buildings with CLT panels. *Sustainability*, **13**, p. 1188, 2021. DOI: 10.3390/su13031188.

[30] Margani, G., Evola, G., Tardo, C. & Marino, E.M., Energy, seismic, and architectural renovation of RC framed buildings with prefabricated timber panels. *Sustainability*, **12**, p. 4845, 2020. DOI: 10.3390/su12124845.

[31] Frunzio, G., di Gennaro, L., Massaro, L. & D'Angelo, F., The CLT panels in structural restoration: Characteristics and technical regulations. *SAHC 2020*, 2021.

[32] Izzi, M., Casagrande, D., Bezzi, S., Pasca, D., Follesa, M. & Tomasi, R., Seismic behaviour of cross-laminated timber structures: A state-of-the-art review. *Engineering Structures*, **170**, pp. 42–52, 2018. DOI: 10.1016/J.ENGSTRUCT.2018.05.060.

[33] Sun, X., He, M. & Li, Z., Novel engineered wood and bamboo composites for structural applications: State-of-art of manufacturing technology and mechanical performance evaluation. *Construction and Building Materials*, **249**, 2020. DOI: 10.1016/j.conbuildmat.2020.118751.

[34] Palermo, A., Pampanin, S., Buchanan, A.H. & Newcombe, M.P., Seismic design of multi-storey buildings using laminated veneer lumber (LVL). *NZ Society of Earthquake Engineering, Annual National Conference*, Wairakei, New Zealand, 2005.

[35] Granello, G., Palermo, A., Pampanin, S., Pei, S. & van de Lindt, J., Pres-lam buildings: State-of-the-art. *Journal of Structural Engineering*, **146**, 04020085, 2020. DOI: 10.1061/(ASCE)ST.1943-541X.0002603.

[36] Hashemi, A. & Quenneville, P., Seismic performance of timber structures using rocking walls with low damage hold-down connectors. *Structures*, **27**, pp. 274–284, 2020. DOI: 10.1016/J.ISTRUC.2020.05.050.

[37] Ganey, R. et al., Experimental investigation of self-centering cross-laminated timber walls. *Journal of Structural Engineering*, **143**, 04017135, 2017. DOI: 10.1061/(ASCE)ST.1943-541X.0001877.

[38] Akbas, T. et al., Analytical and experimental lateral-load response of self-centering posttensioned CLT walls. *Journal of Structural Engineering*, **143**, 04017019, 2017. DOI: 10.1061/(ASCE)ST.1943-541X.0001733.

[39] Pilon, D.S., Palermo, A., Sarti, F. & Salenikovich, A., Benefits of multiple rocking segments for CLT and LVL pres-lam wall systems. *Soil Dynamics and Earthquake Engineering*, **117**, pp. 234–244, 2019. DOI: 10.1016/j.soildyn.2018.11.026.

[40] Sandoli, A. & Calderoni, B., The rolling shear influence on the out-of-plane behavior of CLT panels: A comparative analysis. *Buildings*, **10**, p. 42, 2020. DOI: 10.3390/buildings10030042.

[41] Sandoli, A. & Calderoni, B., The rolling shear influence on the out-of-plane behavior of CLT panels: A comparative analysis. *Buildings*, **10**, p. 42, 2020. DOI: 10.3390/buildings10030042.

[42] Sandoli, A., D'Ambra, C., Ceraldi, C., Calderoni, B. & Prota, A., Role of perpendicular to grain compression properties on the seismic behaviour of CLT walls. *Journal of Building Engineering*, **34**, 2021. DOI: 10.1016/J.JOBE.2020.101889.

[43] Sandoli, A., D'Ambra, C., Ceraldi, C., Calderoni, B. & Prota, A., Sustainable cross-laminated timber structures in a seismic area: Overview and future trends. *Applied Sciences*, **11**, p. 2078, 2021. DOI: 10.3390/app11052078.

[44] Longarini, N., Crespi, P. & Franchi, A., Ronca, P., Scamardo, M.A., & Giordano, N., Dissipative cross lam roof structure for seismic restoration of historical churches. *10th IMC International Masonry Conference*, 2018.

[45] Riccadonna, D., Giongo, I., Schiro, G., Rizzi, E. & Parisi, M.A., Experimental shear testing of timber-masonry dry connections for the seismic retrofit of unreinforced masonry shear walls. *Construction and Building Materials*, **211**, pp. 52–72, 2019.

[46] Salgado, R.A. & Guner, S., Characterization of the out-of-plane behavior of CLT panel connections. *Engineering Structures*, **229**, 111596, 2021. DOI: 10.1016/j.engstruct.2020.111596.

[47] Sustersic, I. & Dujic, B., Seismic shaking table testing of a reinforced concrete frame with masonry infill strengthened with cross laminated timber panels. World Conference of Timber Engineering, 2014.

HEALTHY BUILDINGS: A COMBINATION OF NATURE AND TECHNOLOGY CAN SECURE A SUSTAINABLE FUTURE

BRANDON FUENTES
Department of Architecture, University of Hartford, USA

ABSTRACT

Can buildings be designed to provide people who work in close proximity to each other with physical and emotional comfort or safety, while minimizing the risk of increased viral infection? COVID-19 had a major impact on our country with increased infection rates/deaths, mask mandates, and building occupancy limitations. With the vaccine, mandates were relaxed, including building occupancy levels. Despite this progress, many people are still anxious to work in close proximity to each other in public buildings. This paper seeks to address how healthy building ideals can set the precedent for future public buildings. Healthy buildings are key to bringing back full access to positive communal interaction. Healthy buildings provide for human comfort via advanced building systems, sustainable design thinking, and the utilization low emissive natural materials. Technical innovation and mechanical stability, focusing on indoor air quality and low emissive materials, guarantee occupant safety, while safeguarding a sustainable future. The researcher's proposition was explored through a graduate design thesis project. The building type utilized was a laboratory research center for contagious diseases. Pfizer, Moderna, and the Center for Virology and Vaccine Research were studied as precedents for their current practices related to viral infectious research. Drawing from nature, the geometry and molecular makeup of the atom was used to derive the building form. The design process focused on a combination of natural flora and sustainable technology to eliminate social and psychological fears, while facilitating the required security levels. The results are presented as both an architectural design proposition and related technical and statistical data for a net-zero design. The paper's contention is that research design proposals of this nature help promote healthy building ideals, while ensuring a healthy sustainable future.

Keywords: healthy buildings, critical regionalism, post 9-11 theory, occupancy levels, comfort, safety, connection, proximity.

1 INTRODUCTION

The COVID-19 pandemic has had a major impact on our country with increased infection rates/deaths, brought mask mandates for public settings, and building occupancy limitations. With the rollout of the vaccine, mandates were relaxed including building occupancy levels. Despite all of this, people are still afraid to work in office buildings and return to occupancy levels. So, the question now is, how can buildings be designed to provide people with the comfort and safety of being indoors again while minimizing the risk of infection? My position is to utilize a research center for contagious diseases specifically focusing on viruses to embody the ideals of this question. With this facility, researchers will be able to study and develop more effective vaccines/treatments in a safe controlled environment. Ultimately setting a precedent for what research facilities and buildings in general can become designing with the ideals of a healthy building. Healthy buildings are the key to bringing back the comfort of being indoors. Specifically creating healthy indoor environments through indoor air quality and low emissive materials. Addressing these key elements within the building design can address this question directly while providing the region and the world with a solution to further our understanding of viral infectious diseases and fight against them.

WIT Transactions on The Built Environment, Vol 210, © 2022 WIT Press
www.witpress.com, ISSN 1743-3509 (on-line)
doi:10.2495/ARC220201

2 INSPIRATIONAL QUOTES

"I pictured myself as a virus or a cancer cell and tried to sense what it would be like" [1]. There has always been viruses and diseases spreading around from person to person for centuries. Many people have gotten sick and even died from contracting something as small as a cold in the early years of human civilization. It has not been until recently that early forms of vaccinations were invented to help fight against viral pathogens and diseases. Due to vaccines being so new, there was limited understanding on how they worked and helped protect the human body from infection rather than infecting it. The first vaccine was invented in 1800 by Benjamin Waterhouse and has evolved over time with many other medical professionals trying to help develop more vaccines for other viruses and diseases. This includes developing vaccines for polio just as Jonas Salk successfully did. Jonas Salk was the first American virologist to develop the first successful polio vaccines. The quote above shows how he put himself in the shoes of the virus and diseases to see if he could experience what it is like as these cells. Jonas Salk's mindset of trying to sense what it is like to be viral infectious cells can be utilized for many different applications including aiding in the design of a viral infectious research center.

"Sustainable design is a collective process whereby the built environment achieves ecologic balance in new and retrofit construction toward the long-term viability and humanization of architecture" [2]. The integration of architecture into the surrounding landscape has been a big motive in the past few years and is rapidly becoming the norm in the architecture and construction world. There has been a push for designing buildings as "healthy buildings" increasing indoor environmental quality, sense of place, and well-being. Providing a comfortable and safe place to occupy and conduct daily activities is important when it comes to "healthy buildings". With the presence of the COVID-19 pandemic, "healthy buildings" are in high demand right now and there are not a lot of buildings that have the capabilities of becoming or classifying as a "healthy building". People are afraid to spend time indoors and would rather be outdoors because they feel safer. Integrating the program of viral infectious research with the aspirations and goals of becoming a "healthy building", keeping this quote in mind will help guide what a "healthy building" strives to achieve while mitigating the fears of a facility like this, creating a sense of comfort, and setting a precedent of what buildings should strive to be.

3 BUILDING TYPE

The purpose of current viral infectious research centers is to research and educate people about viruses and diseases while also protecting people from health threats. Developing methods and recommendations such as personal protective equipment (PPE) and basic healthy habits can help limit infection. These centers research viable and safe medications, treatment methods, and vaccines for a multitude of viruses and diseases. Expanding the presence of these centers, not only on a regional scale but on a worldwide scale, can limit the risk of viral outbreaks. The three primary areas of research that are vital to perform viral infectious research are neuro-virology/immunology, researching/developing treatments for neurological disorders, Translational Immunology, researching immune system in relation to new concepts for anti-viral therapy, and viral pathogenesis, understanding how viruses can cause disease in humans to create more effective treatments and vaccines. These three areas of research will be the primary focus of this research center. Physical, computer, and data security are taken into consideration for the occupants/research safety and security.

4 PRECEDENT ANALYSIS

Throughout precedent analysis there were five types of precedents analyzed. Research precedents, looking at the building typology providing information on how existing research centers function and how architects responded to those functions through architecture. Building type precedents, learning about how different projects utilized varied materials, building techniques, systems, and structure to inform the design of the building and site relationships. Technical precedents, studying how other projects integrated advanced building systems, sustainable design methods, and building materials. Abstract precedents, studying a non-building object that has nothing to do with architecture but can relate to the typology of the project and inform the design intent. Anti-precedents, determining what this project does not want to be in terms of design, materials, building techniques, what is deemed to be a failure.

4.1 Precedent utilization

Pearl River Pfizer Headquarters Campus: Pfizer's headquarter campus and the company gave me a better understanding on what types of research they are working on, research methods, and most importantly their vaccine research. *The Center for Virology and Vaccine Research:* The research programs and research processes at this hospital helped expand my knowledge on viral research and provide different options when it comes to what viral research methods I will be utilizing. *Moderna Manufacturing Facility:* Analyzing key elements that Moderna utilizes for their research helped me understand what exactly mRNA is and how mRNA pertains to the development of vaccines. *The Lerner Research Institute:* Understanding their research programs, research processes, and the relationship between the building and the courtyard. *Salk Institute:* Analyzing spatial sequence between the laboratories and the central gathering space, designing for natural light, and open space. Natural lighting was important for productivity in my research center and how to design for natural light utilizing light wells, Northern light, and fenestration methods. *Purdue University Bioscience Innovation Building:* Since this building type is not exactly what my building type is, I was able to pick out certain spaces in the program that helped me start formulating a program. Looking at the structural elements of this building, they exactly resemble what my structural elements are going to look like in the form of lateral cross bracing. Analyzing these elements at this scale was beneficial. *Kenwood Interdisciplinary Research Complex:* Studying the intricate facade and fenestration design. It is interesting to look at the facade and see a primary, secondary, and tertiary facade all within one. Also, the fenestration adding horizontality with shading. *Symbiosis University Hospital and Research Centre:* Analyzing envelope design, solar orientation, and passive design methods. These three elements when combined can create an efficient and sustainable performing building. *Bioprocess Innovation Center:* Understanding the relationships between the surrounding landscape and the built form as well as how the axial circulation works within this building in relation to the surrounding landscape. *Kyushu University Biolab:* Analyzing what a basic research lab layout consists of as well as the necessary ductwork that is required for the equipment in the lab. *Viruses/Antibodies:* Looking at the geometry of virus cells and antibodies for conceptual design inspiration to create architectural/scientific relationships within the center. *Science Research Facilities:* This design is baron, undesirable, region specific materials, essentially looking at how this building design is unsuccessful to use as a precedent compared to what this research center wants to achieve.

5 PROGRAM ANALYSIS

The goal of the program for this project is to provide the necessary and supporting spaces to perform viral infectious research in an efficient and successful manner while also providing the opportunity for sustainable design. Before defining the specific spaces within my program, I looked at my precedents to determine what categories of spaces in each program could work with my building type. After looking at all my precedents, I determined that breaking up the program into categories will help define the use of a cluster of spaces within a specific category. This was when I broke down my building type into five main categories and five supporting categories. The program is broken up into five main categories and five supporting categories. The main categories are research, experiment, green space, meeting space, and visitors space. These categories define how the research center will be utilized and designed for researchers. The supporting categories are mechanical, administration, services, support, and building core. These categories define spaces that will help researchers in the research process get answers and the most efficient methods to go about the research. These spaces also provide necessary security and data management of the research center. Within the program analysis, these categories form the first diagram that could become the parti of the design (Fig. 2). The parti could be used creatively or literally depending on design decisions and goals. Main spaces are listed on the periodic table diagram with their square footages and notation. Below that is a space/major components analysis breaking down the program (Fig. 1).

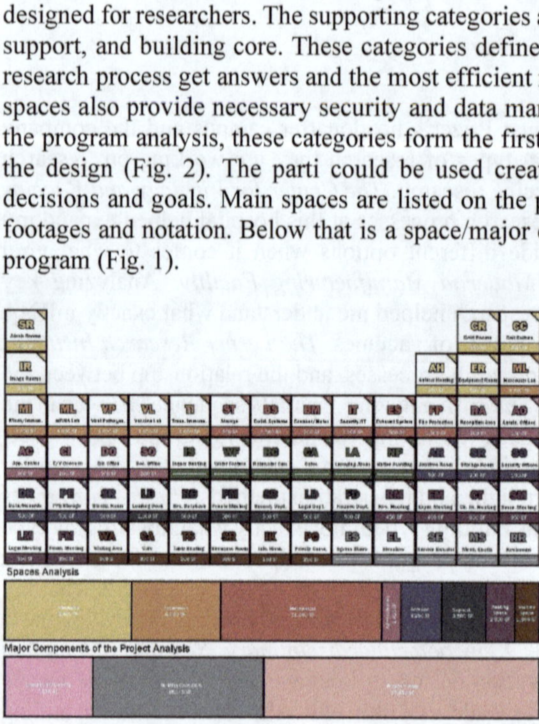

Figure 1: Program periodic table/program scale.

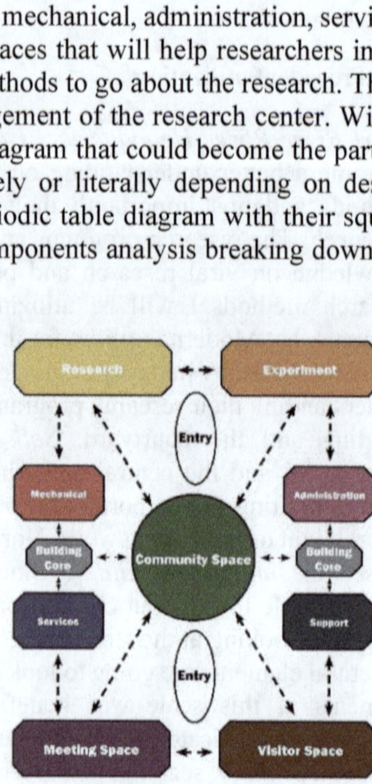

Figure 2: Conceptual parti diagram.

Through this program analysis I have successfully created a program that could work for my chosen building type and benefit the goals and aspirations of what my thesis efforts are striving to be. This program defined the size of the building to be roughly 60,000 SF. After some revisions and program integration, the program developed required double the original square footage with a program square footage requirement of 130,000 SF. The bars signify research/experiment spaces, the spaces between the bars signify support/services, the West side of the building includes meeting/visitors' spaces/administration with mechanical below. From looking at the precedents studied, there were multiple floor plans, space types, and necessary adjacencies referenced to formulate this program. Based on the original program square footage and the revisions of the program, the increase in square footage will benefit the daily spatial needs of researchers.

6 SITE ANALYSIS

The site for this project is located on the eastern coast of the United States in the state of New York. Within New York State, this site is located in Orange County in the town of Wallkill, New York. In the town of Wallkill, there is an existing medical center named Garnet Health Medical Center which specializes in emergency services, healthcare, hospital services, and cancer treatment. Within the Garnet Health campus is the project site just west of the existing medical center. The size of the parcel is nine acres surrounded by trees creating a very secluded parcel which is great for this building type to limit fear of accidental outbreak. This site was previously cleared and flattened but then utilized as a storage location for the dirt that was excavated from the installation of the existing medical center foundation. This has created a big dirt mound that was compacted and had grass planted on the surface to try and make it look natural. Looking at the zoning for the town of Wallkill, New York, there are multiple zones that surround the zone that Garnet Health Medical Center and the project site is in. The zone that these two elements are in is the office and research zone for the town. Since the site and the existing medical center are within the zoning boundaries, it is confirmed that per zoning the building type of a viral infectious research center is feasible in this location. Currently, this parcel does not have a designated street address due to there being no plans for future development currently in this parcel. The closest street address available to locate this parcel was 200 Midway Park Drive which is the parcel across from the project site.

While formulating and compiling the most essential information from the site and the surrounding area, there are a lot of site elements and environmental factors that could contribute or affect the outcome of the project. This graphic is meant to give all the necessary information needed to keep in mind while designing in one place so there is no guessing when designing. The graphic will help make informed design decisions that will benefit the project overall. Smart and effective design decisions based on viable information is key for designing this research center to the best that it can be (Fig. 3).

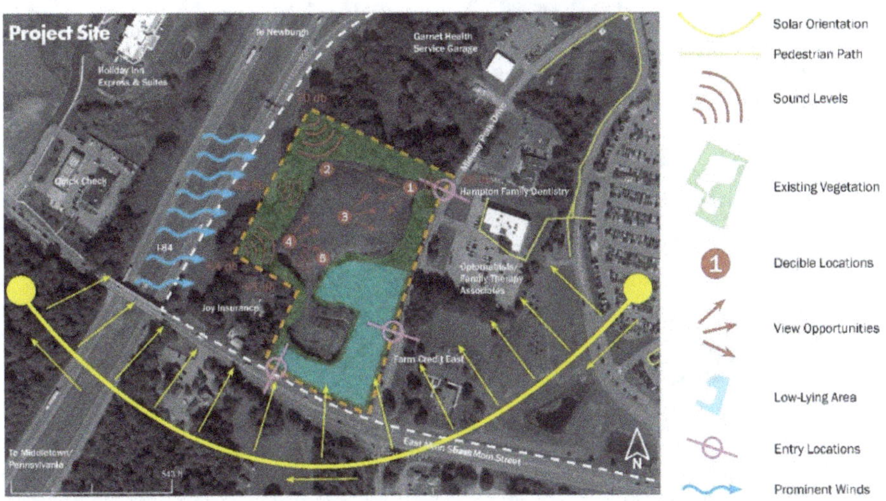

Figure 3: Composite site analysis showing relevant site elements and existing conditions.

7 CONCEPTUAL DESIGN

For concept number one, the idea is to create a functional and productive research environment for all researchers within the facility. There are views from the laboratories, administration, and support areas of the front green space as well as views towards the hospital. The idea of having constant views of the hospital and the green space is to boost the motivation/productivity of the researchers. For concept number two, the idea is to create a highly functional research center with a symmetrical floor plan with four main entrances with access to different program quadrants of the building all surrounding a centralized courtyard. This courtyard provides not only views from the laboratories when researchers are working, it also acts as a space where researchers can get away from research and take a break. For concept number three, the idea is to integrate the building more with the surrounding site and existing topography while providing views towards the existing hospital. The building will be divided into three blocks for three distinct types of research to be performed in these blocks. Laboratories as well as administration and support have views of the rear courtyard as well as the hospital to boost productivity. For concept number four, the idea is to create a building that interlocks together like an antibody creating a strong reference to the natural defense in the human body against viruses and diseases. Since there is a central green space, this concept acts as an active atom with the green space being the nucleus, adjacent spaces being the protons/neutrons, and unique site elements being the electrons. For concept number five, the idea is to reference the Caduceus symbol representing medical practice all over the world creating a sense of hope within the building design as well as the scientists who occupy this building. There are three laboratory sectors based on CVVR research ensuring that the laboratories have an adequate connection/view to nature helping to increase productivity and motivation. For concept number six, the idea is to create a research facility that is more campus like encouraging as much collaboration as possible between researchers with a central collaborative space while also providing research specific laboratories that provide privacy for experiments. An abundance of green space creates an excellent connection with the site and helps the research facility blend in (Fig. 4).

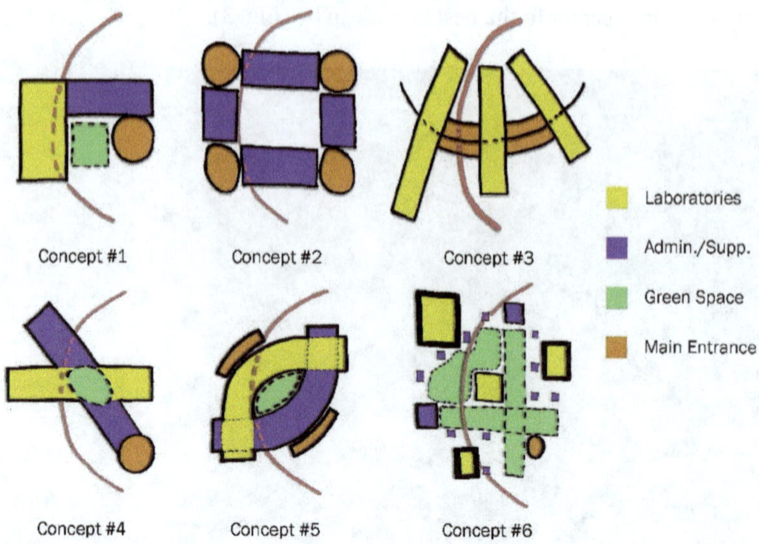

Figure 4: Conceptual design diagrams.

8 SCHEMATIC DESIGN

The concepts that were chosen out of the six from conceptual design for further development were concept #3 and concept #4. Instead of further developing these two concepts separately, they were combined to create three new concepts for the project, concept #7, concept #8, and concept #9 (Fig. 5). For concept number seven, the idea is to also continue exploring concept number three and concept number four together. From further exploration, adding a block towards the back of concept number three created an administrative block. Another block can enclose the green space creating more of a Winter Garden accessible year-round providing a fantastic opportunity for social interaction and entry. For concept number eight, the idea is to continue exploring concept number three as well as combining ideas from concept number four to see if there are any potential opportunities between the two. Taking the bar from concept number four facing East Main Street and putting it behind concept number three allows for an administration block and use of space across the main green space. For concept number nine, the idea is to further explore concept number four to see if there are more opportunities within this concept that can provide a more successful conceptual design. This concept was refined into one main concept but was also explored as two other concepts in terms of building form. Overall, there are three schemes in one concept that can all function the same way. Once these three concepts were developed, concept number seven was further explored to respond to the site analysis accordingly. Concept number seven seemed to have the most potential for the building type and the occupants I will be designing for. Looking at concept number seven, there was something about the rear of the building that did not match the front curve of the building. The rear block of the building was then curved for further exploration of this concept and seeing what the curve in the rear could provide for the building overall. With the refinement and expansion of concept number seven, it helped in the development of the concept for the project. This concept was further explored in terms of curving the administration block to the same radius as the laboratory branches creating the "nucleus" of the laboratory as well as a branding opportunity for the adjacent highway. By creating this "nucleus" promotes the idea of a working atom/cell while also providing researchers an escape from the rigidity of research spaces.

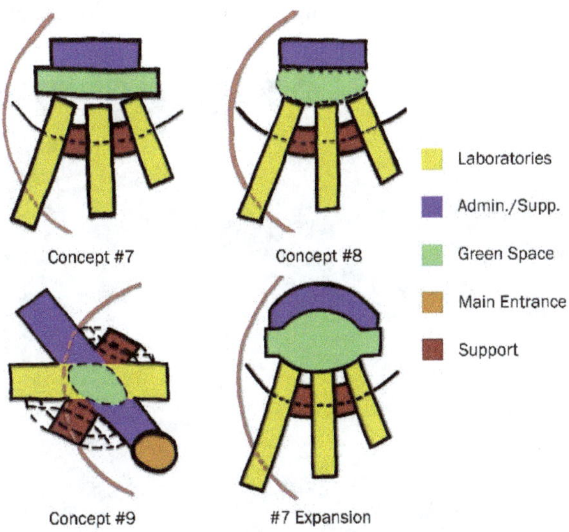

Figure 5: Schematic design diagrams.

9 DESIGN DEVELOPMENT

After thorough refinement and expansion of concept #7, this concept was chosen as the driving force for design development. This concept and building form were developed to show my thoughts and aspirations for this project. Having the concept developed extensively helped in obtaining feedback on the building and design decisions based off a solid concept to back the design. The design development phase included explaining and diagramming the concept of the project, site analysis, proposed floor plans, connections of each branch of the building to different points of the existing medical center adjacent to the site, exploded axonometric drawings, elevations, renderings, and technical drawings such as wall sections to show wall assembly. Once feedback was received, refinements needed to be made to ensure that the building design was showing the overall concept of the design. Sustainable design strategies were analyzed early in the project to be integrated in the design development phase. The feedback provided showed that there needed to be a further exploration on how sustainable design strategies will benefit the overall design based on research and analysis.

Figure 6: Design development rendering of the west elevation facing I-84 highway.

Figure 7: Rendering of chosen concept #7 expansion overall during design development.

10 FINAL DESIGN

Once the schematic design phase was complete, the final development of concept #7 was underway. The final design of this project revolves around the concept of a fully functioning atom with the central atrium space being the nucleus, the laboratories being the protons, the supporting spaces being the neutrons, and the exterior site elements being the electrons. These elements are key for an atom to be an atom and the same thing goes for this building design. Without one or the other, there is no building or "atom". The west side of the building towards the highway has a radial curve to allow for a branding opportunity for the building along the highway (Figs 6 and 8). Creating a more dynamic architectural experience on this side of the building is important to create a separation from the rigidity of viral infectious research within the building. The atrium space acts as the transition space between rigidity and dynamic creating a wonderful place for social interaction within the staff community (Fig. 13). This project aspires to be a LEED Platinum net-zero building (Figs 9 and 14).

Figure 8: final design rendering of the west elevation facing I-84 highway.

Figure 9: Final design rendering of the chosen concept #7 expansion overall.

11 SUSTAINABILITY

Since this building aspires of becoming a LEED Platinum net-zero building there were an abundance of calculations that needed to be performed in order to prove that this building is truly a LEED Platinum net-zero building. Going through the LEED v4.1 BD+C: Healthcare checklist, the project obtained 100 points achieving LEED Platinum certification. There was an energy analysis performed on the building as well as building system metrics done for rainwater collection, collecting 844,302 gallons of water annually, wind turbines, saving $27,000 per year with twelve turbines, geothermal, saving $56,000 annually, and an off-site PV array, saving $149,000 annually. Since the annual energy cost after the energy analysis is $148,376 (Fig. 12), the off-site PV system puts the building at net-zero. Adding a combined savings of $66,000 to the savings from the off-site PV array puts the building at $83,000 in total annual energy cost essentially proving that this building is not only net-zero, but it is in fact a net-positive viral infectious research center. Cooling load calculations were also done to prove that the tight envelope being utilized instead of a loose envelope is decreasing the cooling load of the building essentially limiting cost to the owner and the carbon footprint of the building. A typical laboratory has an air changes per hour rate of 15. Between fan coil units and fume hoods, this laboratory achieves 22 ACPH. Daylighting was also analyzed for fenestration placement to increase sDA and limit ASE (Figs 10 and 11).

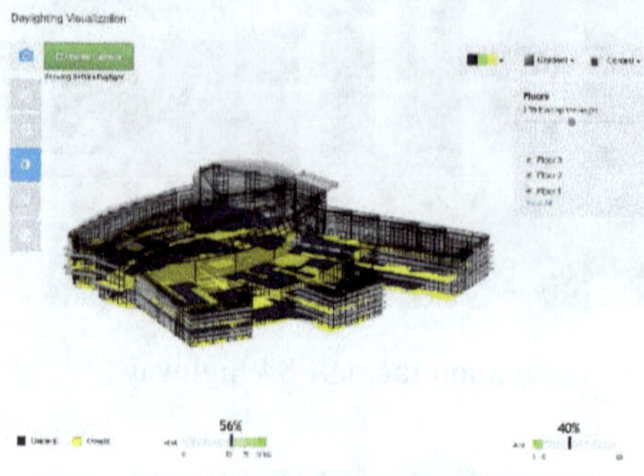

Figure 10: Daylighting with shades.

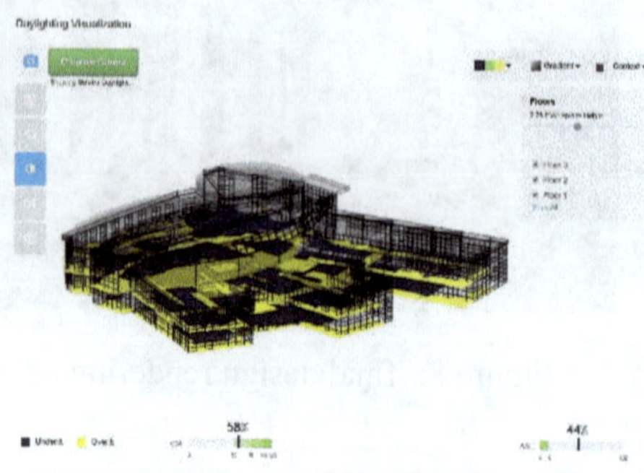

Figure 11: Daylighting without shades.

		HVAC System Type	AHU Design Airflow	Cooling Equipment ...	EUI	Annual Energy Cost
Baseline Concept 130,076 ft²		VAV - 100% OA Central Plant	270,138 (cfm)	1,758.3 ton	170 kBTU/ft²/yr	$665,790
Fan Coil Units		Fan Coil Units and Central Plant	118,545 — 56%	1,088.8 ⌄ 38%	171 ^ 0%	$699,001 ^ 4%
Fan Coil Units, Exterior Ve...		Fan Coil Units and Central Plant	118,545 — 56%	1,059.1 ⌄ 39%	172 ^ 1%	$690,438 ^ 3%
Fan Coil Units, Exterior Ve...		Fan Coil Units and Central Plant	118,545 — 56%	1,037.6 ⌄ 40%	167 ⌄ 1%	$681,908 ^ 2%
Fan Coil Units, Exterior Ve...		Fan Coil Units and Central Plant	118,545 — 56%	1,043.9 ⌄ 40%	158 ⌄ 7%	$672,893 ^ 1%
Fan Coil Units, Exterior Ve...		Fan Coil Units and Central Plant	118,545 — 56%	862.5 ⌄ 50%	141 ⌄ 17%	$558,849 ⌄ 16%
Fan Coil Units, Exterior Ve		Fan Coil Units and Central Plant	118,545 — 56%	1,047.6 ⌄ 40%	115 ⌄ 32%	$477,823 ⌄ 28%
Fan Coil Units, Exterior Ve		Fan Coil Units and Central Plant	118,577 ⌄ 56%	862.9 ⌄ 50%	141 ⌄ 17%	$559,168 ⌄ 16%
Fan Coil Units, Exterior Ve		Fan Coil Units and Central Plant	118,545 ⌄ 56%	1,047.5 ⌄ 40%	92 ⌄ 45%	$338,327 ⌄ 49%
Fan Coil Units, Exterior Ve		Fan Coil Units and Central Plant	118,545 — 56%	1,047.3 ⌄ 40%	61 ⌄ 64%	$148,376 ⌄ 77%

Figure 12: Energy analysis through multiple iterations of different design decisions.

12 CONCLUSION

There has been an extensive amount of research conducted on topics relating to viral infectious research. Throughout this thesis effort, there were deep dives into understanding the medical and research processes in conducting such research as well as the important terminology that goes with it. Analyzing different methods of developing vaccines and fighting against viral infectious diseases expanded the knowledge about how this type of research is conducted and how it works within the human body. Combining the analysis on viral infectious research and the analysis on the building/site side of the research as well as the aspirations of becoming a precedent for "healthy buildings", this research helped direct this thesis effort and ultimately created an amazing design response for the world.

Figure 13: Rendering of V.I.R.C. atrium space.

Figure 14: Rendering of V.I.R.C. researcher entrance.

www.witpress.com, ISSN 1743-3509 (on-line)

ACKNOWLEDGEMENTS

The author would like to thank Daniel Davis, Mark Hopper, and Theodore Sawruk.

REFERENCES

[1] Xplore, Jonas Salk quotes. BrainyQuote. https://www.brainyquote.com/quotes/jonas_salk_390749. Accessed on 15 Nov. 2021.
[2] Loftness, V., Hakkinen, B., Adan, O. & Nevalainen, A., 2007.

Author index

www.ingramcontent.com/pod-product-compliance
Lightning Source LLC
LaVergne TN
LVHW062228200926
848484LV00006B/1501
* 9 7 8 1 7 8 4 6 6 4 7 3 2 *